RODALE'S
COMPLETE
BOOK OF
HOME
FREEZING

by Marilyn Hodges
and the Rodale Test Kitchen staff

 Rodale Press, Emmaus, Pennsylvania

Printed in the United States of America on recycled paper containing a high percentage of de-inked fiber.

Book design by Anita Noble
Illustrations by Barbara Field
Art direction by Karen A. Schell

Library of Congress Cataloging in Publication Data

Hodges, Marilyn.
    Rodale's complete book of home freezing.

    Includes index.
    1. Food, Frozen. I. Title.
TX610.H63 1984   641.4'53   84-11573

ISBN 0-87857-514-6   hardcover
ISBN 0-87857-525-1   paperback

2 4 6 8 10 9 7 5 3 1 hardcover
2 4 6 8 10 9 7 5 3 1 paperback

CONTENTS

ACKNOWLEDGMENTS

Several of the research and testing facilities of Rodale Press were called upon in the creation of this book. Chief among them were the services of the Rodale Test Kitchen, whose staff developed all but a handful of the recipes and tested every last one of them. Their task was to supply imaginative recipes that would freeze well or use foods commonly on hand in the freezer. This they did, producing food that was good tasting, well spiced, and right and ready for the freezer.

Linda C. Gilbert oversaw the selection, development and testing of these recipes. Debra Deis did extensive basic research on freezing specific foods and coordinated the development of new recipes. Marie Harrington coordinated the testing of recipes, as well as developing and testing recipes herself. I am very indebted to them for the knowledge and enthusiasm they brought to this book, and for their careful reading of the text.

Anita Hirsch worked with Rhonda Diehl, JoAnn Coponi, Marilen Reed and Karen Haas to test a large quantity of recipes. Lynn M. Cohen also tested recipes, and Michael Stoner developed recipes. Susan Pearson, Susan Burwell and Ann Sheridan both developed and tested recipes, and Nancy J. Ayers and Vicki Mattern developed, tested and revised recipes. Their work was invariably produced on time and with highly professional results.

Anne Halpin, the book's editor, gave it its final form with a very sure, skilled editorial hand. Bobbie Hartranft did a great deal of fact checking with a great deal of speed. Jayne Tuttle, Dorothy Smickley and Barb Erich expertly typed a long manuscript.

I am grateful to all these people, and to many others who shared their freezer triumphs and disasters with me. In particular, Julie Mayers, whose mother's cat jumped into the freezer and had the whole household hunting for it. And Amy Cohen, who traveled home across three states with a frozen brisket from her mother-in-law, only to unwrap it at last and discover it was a coffee cake.

INTRODUCTION

There's nothing complicated about a freezer. It only does one thing. It fights off the tendency of food to self-destruct by keeping it very cold. That simple function can give you enormous flexibility in the kitchen. Here's how: Use your big, white freezer like the warehouse that it really is. Stock it so that it functions like a whole string of food stores right in your home.

It can be a butcher shop, where you store meat that you bought in bulk at considerable savings. It's your poultry market, where you keep all the chicken parts you bought on sale, and the chicken creations you've cooked in quantity to thaw later. It's your greengrocer, year-round home to the summer bounty fresh plucked from garden, farm and market. In your freezer you can hold onto this fleeting produce—cooked in pies, made into jams and juices, in casseroles, or by the bagful. Your freezer is a fish market, too, from which you can extract the catch from fishing expeditions long after that fish's season has passed.

Get the idea? Instead of stashing food randomly in the freezer, you can use it to control the cost and quality of your food supply. You know how fresh the food is, you know what's in it, you control its availability, and you eliminate waste.

Worthy goals—but how do you start? First, freeze what is worth freezing. Give priority to perishables—meat and fish, fruit and vegetables. Catch the foods that come and go quickly—broccoli, strawberries, salmon—that are abundant and cheap in their season. Freeze the things that you can keep no other way, and then savor them out of season. After all, the next best thing to the pleasures of eating food in season is enjoying it out of season.

How about a December meal of fish caught in July prepared with a bit of dill butter made from midsummer's dill crop, along with some ratatouille prepared and frozen from late summer tomatoes, peppers and zucchini? To finish the meal, a few cherries frosty from the freezer to top a winter fruit salad.

And if you grew the tomatoes yourself, or caught the fish on a lucky day, there is also the real pleasure of eating food with which you're so intimately connected. A feeling of self-reliance can take many forms. Preserving good foods is surely one of them.

And there's more. Beyond freezing perishables, you can use the freezer as a granary, holding all the raw ingredients for grain and pasta dishes. Use it as a bakery, tapping it for breads and cakes and muffins, for dough and batter, and for pie crusts. As a party caterer, your freezer can hold the fixings for small or large gatherings and save you last-minute work. Or it can provide oven-ready meals for days when you don't have time to cook. With this range of food in your freezer, you can have the convenience of *good* food without sacrificing the benefits of *whole* food. You do not have to resort to salty and expensive convenience foods as the price of having quick meals and snacks.

Does all this sound like a lot of work? Too much work for a busy person with a freezer that is supposed to make life easier? Preparing food for the freezer is work, admittedly, but then it is the nature of food to require some sort of human intervention. Almost any food you can name needs some preliminary handling to make it palatable. What are you doing in the kitchen if not trimming, washing, chopping, slicing, heating or chilling foods that have already been hung or milled or pasteurized or dried or plucked or skinned?

Few, indeed, are the foods that you can pop straight into your mouth, and few also are the foods you can pop straight into your freezer. Unfortunately, nothing comes out of the freezer any better than it went in. If you use a freezer as the final resting place for leftovers, that's all it will yield. But if you attend to its care and feeding, it will save you many trips to the market, and it will feed you when you don't feel like cooking.

If you want your freezer to perform productively (and at a cost to run of about 210 dollars a year, it ought to be productive), you have to *use* it. Put things in often, take things out often. Keep the food supply moving and changing. In summer, feed your freezer a diet of fruit and vegetables and fresh-caught fish. In fall, add pork or a lamb or a side of beef bought in bulk. In winter, make soups in quantity and freeze them in small portions. Then use what you froze—first in, first out—and make way for more.

What's going on in your freezer right now? Is it in tune with the seasons? Is it a neat and organized repository of good food, or are the goodies that went in warm, luscious and good enough to save now lying there in terminal obscurity?

When you take food out of the freezer, is it a solid block of colorless, undifferentiated glop—a mess of crystals, slightly risen out of a cracked container? When you thaw it, do you suddenly realize you've taken out chili when you meant to take out spaghetti sauce? When you heat it up, does it smell slightly suspect, and no one wants to eat it? Is it greeted at the table with wails of "Did you *freeze* this?"

If these events sound familiar, it may be that you've assumed that freezing is so easy that you don't have to know much about it. After all, foods just about freeze themselves. True, freezing is the easiest way to preserve food, but a little knowledge is still a useful thing. That's what this book is meant to give you.

For starters, you ought to know that, in every case, freezing changes food. Nothing you freeze comes out of the freezer exactly the way you put it in, and if you are going to embark on using your freezer intensively, as we hope you are, you ought to know what to expect. Some foods do well in the freezer, others shouldn't be frozen at all, and some must receive special attention first. This book will pass on all that information.

Furthermore, food does not last forever in the freezer. It doesn't spoil so that it becomes dangerous to eat, but it will change or deteriorate in flavor and texture. Storage times vary from a few days for homemade ice cream to a year or more for meat. And there is every variation in between. If you want to retrieve good food from your freezer, you need to know not only how to process and package it, but also when to use it.

Much of this information comes from the work of the Rodale Test Kitchen, where nothing is taken for granted. The staff ask questions and seek out the answers: Is blanching necessary? What's the best way to preserve the taste of fresh fish? What's the best way to freeze a large quantity of zucchini? What are the most economical ways to wrap food for the freezer? How do you retain the maximum nutritional value in foods that are frozen? The Rodale Test Kitchen staff cook, too. They've checked every recipe in this book, and all of them conform to Rodale Press guidelines. The recipes contain only fresh, unprocessed ingredients, whole grains, no salt, no sugar, no alcohol and no white flour.

The recipes in this book will let you cook both for and from your freezer. You will find that many of the recipe titles are marked with a ✳ next to the title. That symbol indicates that the recipe can be frozen. The introduction to the recipe will tell you how long you can freeze it. Recipes not marked with the ✳ can be made with frozen ingredients. Wherever possible, for the recipes that freeze we have given yields by volume (such as 1 quart, 1 cup) as well as number of servings, to make container selection simpler.

Because the Test Kitchen staff have tested all these recipes, and many of us involved in the book have tasted them, lots of interesting and helpful comments have gotten into our files along with the recipes. We've included these comments in the book, and you'll find many of the recipes contain variations, serving suggestions and other bits of information that we hope will increase the enjoyment and utility of this book for you.

Finally, a word about freshness. The very thought of freezing food leaves some people cold. Fresh, or not at all, they believe. But, when you think about it, the art of freezing food requires an intense appreciation for fresh food. That is what you are trying to preserve—freshness—so that you can experience it when that truly fresh food is inaccessible. Therefore, freeze only perfectly fresh food. Otherwise the point of freezing is lost. Go out of your way to find the freshest and best-quality produce in season. Buy perfect raspberries, the tenderest peas. Foods that are ordinary and abundant in season are a special pleasure four months later. Use the technological advantage offered by your freezer to prepare food in one season and consume it in another.

In the chapters ahead, you will find the makings of a great many healthful, colorful and satisfying meals. They require only that you use your freezer to keep a hearty stock of good food in good condition. The pages that follow tell how.

PART I

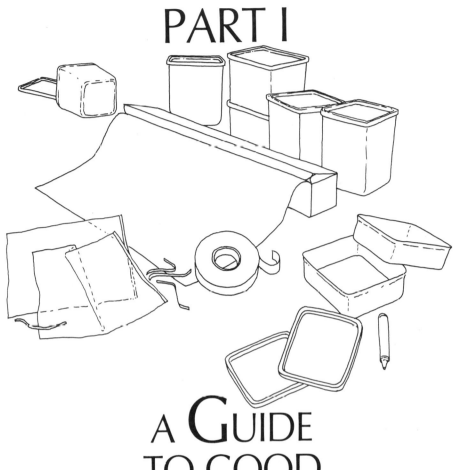

A GUIDE
TO GOOD
FREEZING

CHAPTER 1

FREEZER BASICS

There are two kinds of freezers: chest and upright. Chest freezers, which are always called waist high (though your particular waist may be a lot higher or lower), have a lid that hinges up. Uprights look, and open, like refrigerators.

Chest freezers are cheaper than uprights. They are popular in rural and semirural areas, where there is often more floor space to house their larger bulk. Their design also makes them slightly more energy-efficient; when you open the lid, cold air does not rush out with quite the same vengeance as it does from an upright. When you open an upright freezer, the colder air at the bottom wants out and warm air outside wants in. This exchange makes the compressor run more frequently. In a chest freezer, the cold air at the bottom stays put. With less humid air entering to build up frost, chest freezers need defrosting less frequently, which is lucky because there is no such thing as a self-defrosting chest freezer. You are the self that defrosts it.

If space is tight, you may have to choose an upright simply because it takes up less floor space. If it is not cheaper to buy or run, you will at least find it more convenient to use. The shelves help you organize your storage and then see the food once it is stored. Even if your chest freezer has baskets, and it certainly should, food will tend to jumble up in it. Bending over into the freezer in search of some inaccessible goody may put your head in decisive contact with a rock-hard chicken. With an upright freezer, it may be your foot that obstructs that rock-hard chicken in its quest for freedom. These things happen.

Chest freezers, on the whole, are better at maintaining the all-important 0°F temperature than are uprights, but they, too, have their warm spots. Chests tend to run slightly warmer, usually a few degrees at most, at the top and in the middle of the freezer space. In uprights, the door storage tends to run a bit warmer.

Separate freezers, nevertheless, are far more reliable for long-term storage than is the freezer that is connected to the top, bottom or side of your refrigerator. Though these small freezers do get down to 0°F, you can't rely on them to hold that temperature constantly. A few degrees above zero can cut the storage life of food in half, and these freezers routinely record warm spots of 5° or 6°F, and sometimes as high as 10°F in the door shelves. Their temperature fluctuates with the kitchen temperature and with their more frequent opening and closing. Adjusting a freezer to keep it at 0°F can drop the refrigerator section below 37°F, which is the optimum temperature there. The only answer to the need to keep food on ice for up to a year is to buy a separate chest or upright freezer.

What Size Is Best?

The presence of a freezer in your house can stimulate you to do more gardening, more advance food preparation and more efficient shopping than you

Freezers come in a variety of styles, including small chest, chest and upright.

may have anticipated. There is all that cold space just waiting to hold your pies and produce, your chops and chicken. When you choose a freezer, therefore, be realistic about the amount of time and energy you have for freezing, but also throw in a small measure of optimism for the amount of time and energy you may develop later on. Remember that it's not so much the amount of energy a freezer uses that should influence your choice as it is a question of the amount of energy *you* have to use a freezer efficiently.

For some people, however, the tendency is to get very optimistic about using a freezer, only to have interest wane later on. Before you buy a freezer that is too big—a common mistake—assess the way you manage the freezer space in your fridge. It's a fair guide to how diligently you'll work at keeping a separate freezer full.

You don't want to have more freezer space than you will use, because to operate efficiently your freezer will have to be nearly full all the time. That's because freezing air costs more money than freezing food; it's harder to do.

Chest freezers come in sizes from 5.1 to 28 cubic feet. Uprights are available from 5.2 to 31.1 cubic feet. To give customers a rough idea of freezer storage capacity, the appliance industry has settled on 1 cubic foot as the equivalent of 35 pounds of food. This figure is based on an arbitrary selection of frozen foods of varying densities.

A chest freezer of 5 to 10 cubic feet is a good choice for people who just want to expand their usable freezer space but who don't garden or buy in bulk or cook ahead often. City folks and apartment dwellers tight on space buy this size, squeeze it into a corner and plunk a plant on top.

If your family is small, you do some vegetable gardening and buy food in bulk occasionally, figure on allotting about 3 cubic feet of space per person and buy a chest or upright that has an 8- to 12-cubic-foot capacity.

If you intend to spend the time to plan meals months ahead, freeze some garden produce to enjoy over the winter, buy some of your meat in bulk and precook meals for easy reheating on busy nights, then buy a freezer that holds 10 to 15 cubic feet.

The most popular size freezer on the market is the 16- to 18-cubic-foot-capacity one. This size suits families who have a strong commitment to freezing. That is, about half the food they consume is frozen because they grow lots of fruits and vegetables and often buy foods in bulk when they're in season or take advantage of supermarket specials.

Some people prefer two 10-cubic-foot freezers to one 20-cubic-foot freezer. They keep one freezer accessible for short-term storage, while the other holds large cuts of meat and other long-term supplies. The double system makes it possible to combine the contents into one freezer when stocks run low in spring, and unplug the other. Loading the contents of both freezers into one also makes defrosting easier.

Freezers between 18 and 25 cubic feet are for the most ambitious gardeners and cooks, and for families who wish to share freezing space. Families who depend on their freezers a great deal should consider about 7 cubic feet of space per person about right.

Where Should I Put It?

If you have any choice in the matter, pick a spot out of direct sunlight, in a well-ventilated room near the kitchen where temperatures run between 50° and 65°F for your freezer. Don't assume that keeping your freezer in a cold place will make it run more efficiently. Though it seems odd, freezers are designed to cope with temperatures over 100°F, but not with temperatures below freezing. In a cold climate, therefore, keeping the freezer in an unheated garage is not a good idea.

Chest freezers tend to end up out of sight somewhere in the basement, an enclosed back porch, an attached garage or a utility room. They take up so much floor space that they are usually relegated to parts of the house that are suitably cool but unsuitably distant. Upright freezers have a better chance of taking up residence near the kitchen, but room temperatures and passersby peeking in more often can combine to make the freezer work harder there. The great advantage of keeping the freezer near the kitchen is that, like the appliances you keep on the countertop, what is readily accessible is most often used.

Consider humidity, too. A humid environment encourages the accumulation of frost, which then holds heat in the freezer by preventing it from being absorbed by the condenser coils. The more frost that builds up, the less freezer space you have.

Another point, obvious but sometimes overlooked, is that you should figure out where the freezer will go *before* you buy it. Measure the space you have available, and also the doors, corners and stairwells the freezer will have to negotiate to get there. Remember to add another 2 to 4 inches at the back or sides of the freezer to vent heat from the condenser coils, and to allow you space to

---

Upright freezer doors are hinged on the right, and for several reasons you have no choice in the matter. Their built-in locks make door placement difficult to change, for one thing. Also, since freezers aren't usually in the kitchen, their door placement is considered less important. And you don't open the door that much anyway.

---

clean behind it. If you're buying an upright, it will be hinged on the right and will also require clearance for the door to swing open at least 90 degrees to accommodate its removable shelves or sliding storage bin.

When you are installing the freezer in its perfectly premeasured place, adjust the legs if it's an upright so that the door will close by itself. That is, tilt the freezer back slightly. If your chest freezer is taking up residence in a basement that tends to be wet, mount it on 2 × 4 blocks to raise it above flood level.

How Much Will It Cost to Run?

There's no question that a well-managed freezer is well worth the expense. While other appliances come and go (where are the trash compactors and electric fondue sets of yesteryear?), freezers continue to grow in popularity every year. Because no other way to preserve food is as simple as freezing it, more than one-third of the homes in America now contain the "big white box," and we buy new ones at the rate of 1 million a year. Of the 80 million homes in America, there are now freezers in 30 million of them.

All that we save in human energy by letting the freezer do the work, however, we pay for with a lavish expenditure of electrical energy. That's the only real drawback of owning a freezer—it costs. In a year your freezer will use more electricity than your radio, TV, dishwasher and washing machine combined. Because energy conservation is a national as well as a personal issue, there is ample reason to use your freezer efficiently, to justify the energy cost of producing it and of running it.

If you are buying a new freezer, the cost of running it will be noted right on the model itself in the form of an EnergyGuide label, and you ought to pay attention to it. This label, an innovation of the Federal Trade Commission in 1980, will tell you the estimated annual energy cost of that model and the range of costs among models of similar capacity.

There is a way to look at the whole spectrum of freezer models available and learn their comparative energy costs at one time without a lot of tiring comparison shopping. This information is published in easily readable form in a booklet called *Consumer Selection Guide for Refrigerators and Freezers*. It's available for 1 dollar from the Association of Home Appliance Manufacturers (AHAM), 20 North Wacker Drive, Chicago, Illinois 60606.

To get a really good deal on a freezer, don't look for the cheapest one. Look for the most energy-efficient one, even though it may cost about 10 percent more than a comparable model of average energy efficiency. Freezers have a long life. The payback period for the initial extra cost will be about two years, and the energy savings will continue for eighteen more years. Consumers, generally, have been slow to realize that buying energy-efficient appliances is highly cost-effective. More than any measures you can institute to save energy once you have a freezer, your greatest savings will come from simply buying an energy-efficient model in the first place.

What if you bought your freezer before the EnergyGuide system was instituted? If you have a pre-1980-model freezer, you can't know exactly how much it costs to run, but it is certainly more than a new model would cost. New freezers have more and better insulation, which is made of polyurethane foam rather than fiberglass. Refrigerant systems have improved, and motors and compressors are more efficient. These improvements have been simple and sensible, and they definitely make freezers better than they used to be.

There are some statistics that can help you gauge the relative efficiency of your older freezer. A freezer made in 1972 consumed an average of 1,460 kilowatt hours (kwh). One made in 1981 consumed an average of 850 kwh. By 1983 freezers were 55 percent more efficient than they were in 1972.

How to Keep Your Freezer at Peak Efficiency

There is not a lot you can do to significantly lower the energy costs of the freezer you have, but the following measures will help keep those costs from escalating unnecessarily.

First, keep the freezer set at 0°F. Lower temperatures hold food longer but do so at such a high energy cost that it makes poor economic sense. Besides, six months to a year is a reasonable length of time to hold food, and that duration is fully within the capacity of a freezer running at 0°F. If the temperature is higher than 0°F, the food will deteriorate far more quickly.

Second, keep it 75 to 80 percent full. Frozen food acts as a thermal mass and helps to keep the temperature down. Keeping air cold takes more energy than keeping food cold. Stock up on bread and other bulky foods when your other supplies are low. Fill gaps in the freezer with milk cartons full of ice, which you can then use in the summer to cool blanched vegetables that are destined for the freezer.

Third, don't open the door often, particularly when hot, humid air can tumble in to replace the cold stuff. The freezer has to work to cool the incoming air, and the food suffers from fluctuating temperatures.

Keep the freezer away from the woodstove or the furnace, in a spot where the ambient room temperature is between 50° and 65°F.

Make sure the freezer is well insulated and well sealed. If the gasket around the door is worn or hardened, it leaves gaps that allow cold air to leak out and warmer air and moisture to slip in and create frost. The more frost that builds up, the harder the freezer has to work.

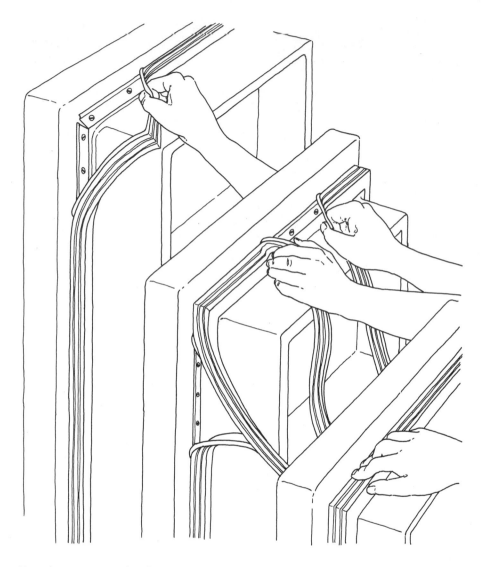

To replace a worn gasket, first loosen but don't remove all the screws. Begin pulling the old gasket from behind the retainer, top, and feed in the new gasket with your other hand, center. Finally, push the bead around the new gasket under the retainer and tighten the screws.

To test the seal on a freezer door, place a 150-watt outdoor floor light inside the cabinet. Direct the light at one length of the gasket at a time with the door closed. The gasket will allow the door to close on the cord. If some light is visible at any point, the gasket is defective.

To replace a worn gasket, loosen the retainer screws, but do not remove them. Pull the gasket out a little bit at a time from behind the retainer. Feed in the new gasket as you remove the old. If you remove the old gasket all at once, the door may warp. With your fingers, push the bead around the new gasket under the retainer. Some freezer doors have a newer gasket with a continuous magnet embedded inside it. To test its effectiveness, look at it to see if it touches the door jamb all the way around the door. If it doesn't, replace it.

Another aspect of good freezer management is simply to use what's in your freezer—don't just save it. Use the food so that the freezer contents are emptied and replaced at least twice a year. The pie that stays in the freezer for eight months is costing a lot more to keep cold than any one of a succession of packages that you store and use.

There is some debate about whether a manual-defrost freezer is cheaper to operate than an automatic-defrost model. Typically, an automatic-defrost model is said to cost about one-third more to run, but there are other factors involved that this figure does not take into consideration. The accumulating frost in a manual-defrost model can insulate the cooling coils from the interior and make the freezer work harder to stay cold. Turning the freezer off to defrost and then on again to re-cool costs money, too. Since the same kilowatt-hour of electricity that costs 3 cents in Seattle costs 15 cents in New York City, and since some manual-defrost freezers are much easier to defrost than others, it becomes a moot point which system is better.

In an area with high kwh costs, a freezer with a flash-defrost button is probably your best bet. You need only remove the food and press a button that activates heating coils inside the walls near the top. The frost melts on the walls; allowing you to lift off the ice in large pieces and be done with the job in half an hour.

How a Freezer Works

The inner workings of a freezer do not create cold air. Rather, they transfer whatever heat there is in the closed box to the outside. Refrigerators and air conditioners do roughly the same thing.

If the workings of your freezer are a mystery to you, it's because they take place in a sealed system tucked inside the sides and around the bottom or the back. There, where you can't see it, a vehicle called a refrigerant is used to convey heat from within the freezer to the outside. Though most people know the refrigerant by the trade name Freon, there are actually several different varieties.

REFRIGERANT CIRCULATING SYSTEM IN A FREEZER

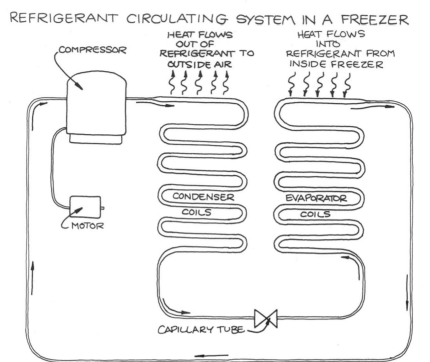

Though home freezers are a twentieth-century phenomenon, they work by applying a principle discovered in the eighteenth century: when a substance changes from a liquid state to a gaseous state, it absorbs heat from its surroundings. Conversely, when a substance changes from a gaseous state to a liquid state, it releases heat to its surroundings. The refrigerant inside a freezer is sometimes in a liquid state and sometimes in a gaseous state, and it behaves accordingly, alternately extracting heat from the air inside the box and releasing heat to the outside.

A motor drives the pump, called a compressor, which forces the refrigerant through the system. The compressor squeezes the refrigerant, in its liquid form, through the capillary tube. As the refrigerant passes through the capillary tube, the pressure on the refrigerant is reduced and it evaporates into a gas.

Passing into a set of coils called the evaporator, the gas extracts the heat from the freezer right through the walls of the evaporator, causing the freezer to get cold. The evaporator coils in a chest freezer may be wrapped around the whole cabinet, inside the insulated walls. In an upright, they are sometimes in the shelves as well.

Still a gas, the refrigerant now flows back to the compressor, where it becomes a high-pressure gas. The compressor then pumps it on to the condenser, another set of coils. Here it condenses back to a liquid, giving off its heat in the process. The heat moves through the walls of the condenser coils into the room. The condenser, which is exposed to the room air, is usually at the back of the freezer but is sometimes underneath it.

When the condenser coils have dissipated the heat, the refrigerant, cold once more, changes back to a liquid and continues on its way around the system again. This endless recycling of the same refrigerant through the sealed system leaves little room for mishap. Surprisingly, if your freezer is ever going to have a problem, it is likely to be in the first year. If no flaw manifests itself by then, you can expect the system to function trouble-free for about twenty more years. The chances are only one in ten that you will ever have to call for repairs.

Freezer Maintenance

If your freezer could talk, you would most want it to tell you when it is getting too warm. After all, the quality of your frozen food is directly linked to the temperature at which it is stored and the extent to which that temperature is consistently maintained. There are two devices that can give you this vital information without your needing to peer inside the freezer and let out the cold air, and it is worth owning both of them.

The first is an inside-outside freezer thermometer. It consists of a sensor that you place inside the freezer and a mercury thermometer that you attach to the outside of the freezer door. The two parts are connected by a long copper wire that runs between the door gaskets without affecting the seal. If the thermometer ever indicates that the temperature inside is not where it ought to be, just fiddle with the freezer's control knob until it's right again. (If

A freezer lasts a long time. Generally, the patience of the freezer's owner will break down before the freezer does. It can happen that you just get tired of looking at the same freezer for twenty years and decide you need to have a new one. If your patience doesn't give out, eventually the compressor will.

you can't get the temperature to be right again, your freezer may be signaling that all is not well in its workings.)

An ordinary freezer thermometer, on the other hand, will not give as accurate a measure. Because you must open the door to read it, it only tells you what the freezer temperature is when the door is open. It will be no help at all if you are worrying your way through some kind of power failure, afraid to open the door but unable to know what's going on inside any other way. That's when a door-mounted thermometer really comes into its own.

The second wise buy is a freezer alarm. You need one if your freezer has no warning system that sounds when the internal temperature is reaching suspicious levels, or even a light that indicates when the power is on. Buy a battery-operated temperature-sensing device that sounds a very loud, long-lasting (days if need be) alarm when the freezer temperature is rising dangerously. The alarm has a sensor that reads the internal air temperature (because the air warms up faster than the food) and is connected to an outside alarm control box by a thin cord that passes through the door seal without letting air escape.

Defrosting

When the ice inside your freezer gets to be between ⅜ and ½ inch thick, it's time to defrost. Depending on your local climate, how often you open the door and how often you scrape ice off the sides as it forms, you'll need to defrost at least once a year and possibly twice.

If you have seasons to choose from, defrost in the dead of winter, when you can leave the food in the ice outside while you chip away at the ice inside. Bury the food in the snow, out of reach of the winter sun. Or, follow the boom-and-bust harvest cycle and defrost in spring when your food stocks are at low ebb.

The day before you plan to defrost, turn the freezer temperature down to give the food an extra shot of cold to help it through the next day's dislocation. On Defrosting Day, pull the plug. Pack the food into picnic coolers, put it in the fridge if there's space, or line boxes with newspaper, add the food and cover it with blankets.

Scrape the frost, but not the finish, using a wooden or plastic implement. Add pans of hot water to speed thawing. Or set up a big electric fan on a card table in front of the freezer. If you're really in a hurry, add an electric heater behind the fan.

When the ice is gone, wipe out the freezer with a wet cloth sprinkled with baking soda. Rinse it well. Dry it thoroughly with a towel, let it air dry for a few minutes, then plug it in again. Run it for ten minutes or so, then retrieve the food and pile it all back in. Congratulate yourself on a job well done.

When the Power Fails

It may well happen: The cat pulls the plug, a storm downs the power lines, or the freezer develops indigestion. Suddenly, you find yourself grappling with freezer meltdown. But you are not as powerless in this situation as you may think.

The first thing you can do is anticipate disaster well in advance and be ready for it. Track down a source of dry ice ahead of time, and keep the phone number taped to the outside of the freezer. Try the Yellow Pages under Dry Ice, or Ice, or Ice Cream Manufacturers, or Refrigeration Suppliers. Your electric company, supermarket, local dairy or fish market may also know sources.

If dry ice sources fail to materialize, track down the location of a freezer-locker plant. Look in the Yellow Pages under Warehouses—Cold Storage or Butchering. Again, keep the phone number taped to the freezer. Also, keep in mind which of your friends have freezers and may be able to babysit for your food in an emergency. Now you're prepared for the worst.

If you suspect that a power failure is imminent, turn the freezer temperature down lower to prolong the thawing process. If you do lose power and you know when it happens, you're in a good position to save your food supply. Don't open the freezer door. Assume that your food will stay frozen for one day if the freezer is less than half full. If the freezer is well-loaded expect it to stay frozen for two days. Your food may stay frozen even longer if the freezer is a big one and packed with meats, which are dense and slow to thaw.

If it looks as if your power failure may be prolonged, buy some dry ice. To arrive at the correct number of pounds to buy, multiply the cubic capacity of your freezer by 2.5. That amount will keep food frozen two to three days in a half-full freezer and three to four days in a fuller freezer.

Dry ice is extremely cold, -110°F. It is so cold that if you touch it without heavy gloves on you will develop frostbite and lose some skin. If you pack it directly around your food, it can cause dehydration. To be safe, put on heavy gloves, heap the food together in a chest freezer and cover it with heavy cardboard or wood. Wrap the dry ice in a lot of newspaper and lay it on the cardboard. Because cold air travels down, you will have to follow this same placement on each shelf of an upright freezer. To make room for the dry ice in a nearly full freezer, remove food you can use for the next several meals.

Because dry ice (carbon dioxide) vaporizes from a solid state to a gas, it won't melt and drip all over your freezer. Instead, it will evaporate. The carbon dioxide will dissipate through the minute cracks in the door gasket (no seal is 100 percent tight). When you are handling the dry ice, provide for some ventilation in the room, avoid inhaling it and don't chip at it with an ice pick. It splinters like glass and can cut you.

The alternative to using dry ice is to trundle your frozen food to a storage locker or a friend's house. Wrap the food for the journey in abundant layers of

(continued on page 18)

REFREEZING—WHEN IS IT SAFE?

FOOD	ICE CRYSTALS STILL INTACT, FOOD STILL COLD (40°F OR LESS)
BAKED GOODS	
Breads, rolls	Refreeze
Unfrosted cakes	Refreeze
Unbaked pies	Refreeze
Baked pies	Refreeze. If made with dairy product in filling, use immediately.
Cream filled cakes & cookies	Use immediately.
DAIRY	
Milk	Use immediately.
Foods containing any form of dairy product	Cook and serve immediately or refreeze if dish was not cooked before initial freezing.
Ice Cream	Eat immediately.
FISH	Cook and serve immediately.
Shellfish	Cook and serve immediately or refrigerate no more than 2 days before using.
FRUIT	
Fresh	Refreeze
Commercially packaged	Questionable. See package directions.
Fruit juice concentrates	Refreeze
GRAINS	
Dried beans	Refreeze
Uncooked rice	Refreeze

THAWED, BUT AT ROOM TEMPERATURE LESS THAN 3 HOURS	THAWED AND AT ROOM TEMPERATURE FOR UNKNOWN TIME
Refreeze, but some drying may have occurred.	Refreeze, but drying may have occurred.
Refreeze, but some drying may have occurred.	Refreeze, but drying may have occurred.
Refreeze, but some drying may have occurred.	Refreeze, but drying may have occurred.
Use immediately. If filling contains dairy product, discard.	Discard
Discard.	Discard.
Discard	Discard
Discard	Discard
Discard	Discard
Cook and serve immediately. If odor is detectable, discard.	Discard
Cook and serve immediately unless odor is apparent.	Discard
Discard unless a high-acid variety.	Discard
Discard	Discard
Refreeze, but flavor and texture may have deteriorated.	Discard, fermentation may cause can to explode.
Refreeze	Refreeze
Refreeze	Refreeze *(continued)*

REFREEZING—WHEN IS IT SAFE? *(continued)*

FOOD	ICE CRYSTALS STILL INTACT, FOOD STILL COLD (40°F OR LESS)
GRAINS—CONTINUED	
Cooked rice	Refreeze, but texture and flavor will have deteriorated.
Uncooked pasta	Refreeze
Cooked pasta	Refreeze, but texture and flavor will have deteriorated.
Flours	Refreeze
Nuts and nut meats	Refreeze
HERBS, SPICES & SAUCES	
Herbs	Refreeze
Spices	Refreeze
Sauces	Refreeze. If they contain any dairy products, cook and use immediately.
MEATS	
Beef, veal, lamb	Refreeze
Pork, fresh	Refreeze
Variety meats (liver, kidney, heart, etc.)	Refrigerate and use within 48 hours. Do not refreeze after cooking.
Casseroles, stews, meat pies, and all combination dishes containing meats	Cook and serve immediately or cook and refreeze. If the dish was cooked before initial freezing, do not refreeze again after cooking.
Commercially packaged casseroles, dinners, etc.	Questionable. See package directions.
POULTRY	Refreeze if considerable amount of ice crystals are apparent.

THAWED, BUT AT ROOM TEMPERATURE LESS THAN 3 HOURS	THAWED AND AT ROOM TEMPERATURE FOR UNKNOWN TIME
Cook and serve immediately.	Discard
Refreeze	Refreeze
Cook and serve only if pasta has not melted together.	Discard
Refreeze	Refreeze, but flavor may have deteriorated.
Refreeze	Refreeze
Refreeze	Can refreeze, but will be very mushy
Refreeze	Refreeze
Use immediately. If they contain any dairy products, discard.	Discard
Cook and serve or cook and refreeze.	Cook and serve immediately. If odor or off-color is detectable, discard.
Cook and serve or cook and refreeze.	Cook and serve immediately. If odor or off-color is detectable, discard.
Cook and serve immediately. If odor or off-color is detectable, discard.	Discard
Cook (or reheat thoroughly) and serve immediately.	Discard
Cook and serve or cook and refreeze. If odor or off-color is detectable, discard.	Discard

(continued)

REFREEZING—WHEN IS IT SAFE? *(continued)*

FOOD	ICE CRYSTALS STILL INTACT, FOOD STILL COLD (40°F OR LESS)
SOUP	Refreeze unless it contains meat, fish, poultry or a milk product. Cook and serve these immediately.
VEGETABLES	
Garden	Refreeze. May have lost some texture and flavor.
Commercially packaged	Questionable. See package directions.

newspaper and blankets, or pack it tightly into insulated cartons.

What do you do if a power failure has come and gone while you were away from home? When you return, your only clue to the event may be your electric clock. If it is reading 4:15 when it's actually 9:30, all is not well. Proceed directly to the freezer. Keep a bag of ice cubes in your freezer specifically for this contingency. If you find them frozen together in a big clump, you'll know that the freezer has thawed them and then refrozen them.

If your ice cubes tell you that thawing has taken place, your next task is to learn the duration of the power failure. With that information in hand, perhaps from a neighbor, you can calculate roughly what degree of thawing took place. If the failure was less than a day or two long and your freezer was reasonably full, you can relax. The food stayed frozen.

But what if you're in the position of having to assess the damage to your frozen food as the result of a longer power failure? You'll be forced to make some decisions based on nothing more than good guesswork. Here are some guidelines to help you decide how much the food has thawed, which foods you can refreeze, which ones to cook immediately and which to throw away. The questions you must answer are: How much has the texture of the food deteriorated? And have harmful bacteria had a chance to become active and affect the food?

The best ways to check are to squeeze wrapped foods without unwrapping them and to take the lids off foods in containers and look inside. Foods that feel quite firm and have lots of ice crystals are okay to refreeze.

Food that has passed the point of being still obviously near frozen is harder to assess. In theory, anything that is still at refrigeration temperatures (no warmer than 40°) and contains some ice crystals is okay to refreeze, with

THAWED, BUT AT ROOM TEMPERATURE LESS THAN 3 HOURS	THAWED AND AT ROOM TEMPERATURE FOR UNKNOWN TIME
Discard	Discard
Cook and serve immediately or cook and refreeze.	Discard
Discard	Discard

certain exceptions. Precooked meals that include meat, fish or poultry should be eaten quickly. The same is true of garden vegetables, variety meats like liver, kidney and heart, fish, ice cream and any other foods that contain cream. None of these foods should go back into the freezer, and neither should any other foods that you feel hesitant about. Better to err on the side of caution.

If the food seems completely defrosted but has been at room temperature for only a couple of hours, the breads and rolls, certain fruits, and fruit juice concentrates can go back into the freezer, though all will have suffered some deterioration. All the raw meats and poultry should be cooked immediately and then refrozen. Everything else is forever barred from the freezer.

If the food is thoroughly defrosted and you don't know how long it has been at room temperature, don't flirt with food poisoning. Brace yourself and pitch it out, with these exceptions: If the raw meats and poultry smell and look okay, cook them up quickly, but don't refreeze them. Don't use the sniff test on anything else, though. Low-acid foods including shellfish, vegetables, cooked foods and dishes containing cream may smell fine even though they are highly susceptible to bacterial growth, and they can do some decidedly un-fine things to your tummy. You can refreeze the bread and rolls, highly acidic fruits and unfrosted cakes, but that's it.

Finally, check your freezer warranty. The only consolation you may have if you find yourself surrounded by dripping bags of discarded goodies is the manufacturer's promise to reimburse you (usually to around a 200-dollar limit) for foods that thaw. But the thawing has to be the result of the freezer breaking down. If the cat pulled the plug, you're out of luck. But you can prevent a recurrence by installing clips that secure the plug in the wall receptacle. Appliance manufacturers sell them.

CHAPTER 2

ABOUT
FROZEN FOODS

Be fond of your freezer. It's kind to your food. First, it provides a safe haven for more foods than any other preservation method. It's a service foods can well use. Take a ripe blueberry. It looks perfect, but there are two things wrong with it from the cook's point of view. For one thing, unlike a chair or a hammer, you can't count on it to be available anytime you want it—you couldn't find it in December, for instance. It has its season and then it's gone. Secondly, when it is in season, it doesn't stay ripe for very long.

The trouble with blueberries, and all other foods, is that they're perishable. Their fate in nature is to decompose and become again the organic compounds from which they originated, returning to the soil to nourish a new crop and becoming part of an endless biological cycle. Enzymes and microorganisms— molds, yeast and bacteria—are the vehicles for breaking down food. If they did not exist, unused food would pile up on the earth. But the presence of enzymes and microorganisms can be a nuisance for us, too.

Given that all food is organic material, one point of the cook's task is to stave off or suspend its natural tendency to decompose. Controlling enzymes is part of the job. Enzymes are protein molecules that act as catalysts. They are responsible for the ripening of food, but they don't know when to stop. They are also responsible for food going beyond ripening to decay. They cause the bananas and green tomatoes on your kitchen counter to ripen and then to become spotted, mushy and finally rotten. While enzymes are at work within, bacteria attack food from the outside. Creepy as it sounds, bacteria reproduce themselves faster than any other life form. It sounds like a science fiction movie.

Enter cold, coming to the cook's rescue and producing a chilling effect on the elements in food that cause it to go bad. Freezing food at 0°F slows down the enzymes and inactivates the molds, bacteria and yeast that are present in food. It doesn't kill them all—only extreme heat and extreme cold can do that—but it sends them into a sound sleep from which only thawing will rouse them. At freezer temperatures microorganisms can't multiply, but between 40° and 140°F they sure can. That's why refrigerators are best kept at a temperature of 37°F.

While freezing is warding off spoilage, it is doing food yet another favor. It preserves it gently, without drastically altering its flavor or texture. All preservation methods alter food, of course, but freezing preserves the widest range of foods with the smallest degree of change. Freezing is the preservation method that gets the prize for keeping food closest to fresh.

Canning processes, on the other hand, control microorganisms and enzymes by a more extreme method. The food and its container are heated to such high temperatures that both are sterilized. But in the process the food is significantly changed, just as it is when it is dried.

Drying food removes from 80 to 90 percent of its water, and along with the water go all the harmful microorganisms, because every life form relies on water. Dried food is easy to store and doesn't take up much space, but it is vastly different from fresh.

Freezing does not sterilize food, but neither does it substantially alter its basic composition during processing. Nor does it significantly change its nutrient content. Here again freezing holds food closest to fresh. In ideal freezer conditions there is negligible loss of vitamins. The reality, of course, is less than ideal—temperature fluctuations happen, freezing takes place slowly, foods sometimes stay too long in the freezer. But even then, food in the freezer holds its own nutritionally.

Now take another look at that imaginary blueberry. The best way you can save it to enjoy out of season is to freeze it. The best way to make it worth saving is to treat it well in the kitchen before it goes into the freezer and after it comes out. Nutrient loss is most severe while the cook is handling the food, not while it is in the freezer. Vitamins leach out of food during cooking, drip out of it on the countertop, get peeled and trimmed from it during cleaning. To best enjoy the substantial benefits of freezing food, follow the guidelines in the next chapters.

Freezing and Thawing,
or Where Did Those Ice Crystals Come From?

It's essential to get your food into the freezer as soon as possible. It won't come out any better than it went in, and the longer you delay, the greater the deterioration that is taking place. All the time you are holding food in the refrigerator or on the counter, it is losing freshness and nutrients, particularly vitamins A and C and folic acid. Here's one graphic estimate of how fast food deteriorates: Beans will last five days at 30°F, three weeks at 20°F, three months at 10°F, and one year at 0°F.

The moral of the story is: Get your food frozen as quickly as possible. This is very important. Your goal is to preserve as much as possible of the original texture of the food (even though freezing will inevitably alter any food somewhat). To do so, you have to minimize the formation of ice crystals, and the longer it takes the food to reach 0°F, the larger the ice crystals will be. Here's why. The water content in food expands as it freezes. As it expands, it forms ice crystals, which push against the food cells, bursting the walls of some, pushing others aside. If food is freezing slowly, it creates an opportunity for large ice crystals to form.

When the food thaws, if the cell walls are not in good shape, they will provide places for the liquid to escape and drip away. When liquid seeps out of food, it destroys its firm texture, flavor and nutrients.

How do you freeze food as quickly as possible? First, place it in the freezer only when it is no longer warm. Put in no more than 2 pounds of meat or fish or 3 pounds of less dense foods like fruit and vegetables at one time per cubic foot of freezer space. Otherwise, the temperature of the freezer will rise with the quantity of warmer food in it, and freezing will take longer. The temperature rise will cause the formation of larger ice crystals in both freezing and already-frozen foods. Fluctuating freezer temperatures are a prime cause of ice-crystal damage.

Put the food in the coldest part of the freezer at first. Spread out the food in a single layer, with space around each package in order to maximize the circulation of cold air. (Never pack the freezer so full that cold air can't circulate well.) In chest freezers, the coldest parts are against the sides. Chests are warmer near the middle and the top. In uprights, put the packages right on the shelves. Uprights are warmest on the upper door shelves and near the front. After 24 hours, move the packages to their long-term positions.

Freeze food fast, but defrost it slowly. The right place to defrost food is either in the refrigerator or in a waterproof container in cold water or under running cold water. Because food that has been frozen has a diminished ability to retain water in its cells, that water seeps out of its cells when it is thawing. The seepage makes the cells less plump and the food less firm. If you thaw food on the counter while you are away from home, rather than in the refrigerator, it may thaw and seep longer than it needs to if you let it reach room temperature.

For example, a chicken defrosting on the counter will drain juices full of B vitamins. In the refrigerator it will hold at below 40°F until you get to it. On the counter it may well reach temperatures at which microorganisms start to flourish before you get home to start supper. As mentioned earlier, the danger zone, where bacterial growth is prolific, occurs at temperatures between 40° and 140°F.

True, you have to think well ahead to thaw food in the refrigerator. The roast that thaws on the counter at a rate of 1 pound every three hours will take six hours per pound in the refrigerator. There are some foods that you can cook from frozen, like soups and sauces, if you allow twice the time and use a lower temperature. Vegetables, except for leafy greens and corn on the cob, should go straight from the freezer to the pot. Other foods, though, like thick cuts of meat, must thaw completely to get even cooking results. Thawing rules vary with each food group, so see the relevant food chapter for more specific information.

A further bit of advice: Use thawed foods quickly. As the temperature in defrosting food rises, so too does the rate of growth of the microorganisms it contains.

What Freezer Coverings Do

As soon as you say "Bye, bye, baby limas, see you in three months," and pop them in the freezer, they will immediately be under siege. Unless you have first wrapped them in a coat of protective armor, your frozen foods will be doing battle with air, moisture, the odors of the other foods, the weight of the packages around them and the tendency of their own odors and juices to want to escape. Only if you have packaged them properly will they emerge from the assault with their flavor and texture intact.

Use food coverings that are nonporous, to prevent flavors and odors from entering or leaving. Coverings should have no taste or odor of their own, either. They should be moisture-proof, too, and seal in a way that excludes as much air as possible. If the air in your freezer comes in contact with your food, it will dry it out. That dehydration, misnamed freezer burn, is the direct consequence of poor packaging. It makes juicy, red steak turn tough and gray and taste like cardboard. When food contracts freezer burn, it's terminal. You might as well throw it out because there's no way to restore the quality.

Freezer packaging also has to be strong and thick because the packages of food are stacked so closely together—peas on top of carrots, berries under pork chops. Jostling, too, can take its toll, as you dig through the pie dough to get to the chives, or upend the turkey to find the croquettes. All this can wear and tear weak packages.

The Best Materials to Use

There are all sorts of materials and containers you can use for freezing food. The discussion that follows will summarize the merits, drawbacks and costs of the most popular freezer coverings.

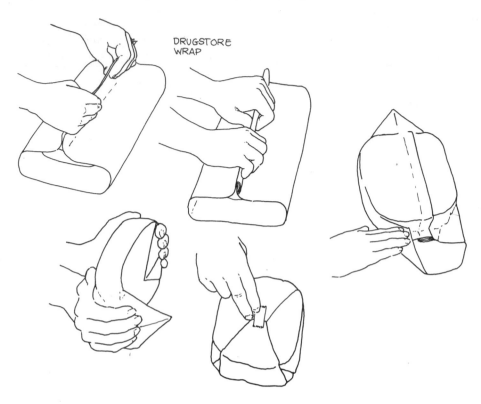

DRUGSTORE
WRAP

For the drugstore wrap, bring the long sides of the wrap to the center over the food and fold over about an inch. Crease the wrap and continue making 1-inch folds until the wrap is close to the food. Press out all the air, fold the ends toward each other to form a point, then fold the points under the package toward each other. Seal all the edges with tape.

Wraps

For flexible packaging buy commercial freezer paper (which is strong paper coated with plastic to retain moisture) or heavy-duty aluminum foil. They cost roughly the same. Though the various brands of each differ in price, they all do a serviceable job of maintaining a moisture/vapor barrier around the food. Buy the least expensive kind of whichever of the two you choose. Don't reuse the foil for long-term freezer storage, however. Heavy duty though it may be, it will inevitably have developed pinholes that hamper its impermeability the second time around. Skip the regular-strength aluminum foil, too. It's not strong enough for the freezer. And pass up the extra-heavy-duty foil. It's stronger than you need, and doesn't justify the added expense.

Not all plastic wraps are suitable for use in the freezer. Buy only the ones that are labeled as freezer wraps. They're usually made of extra-thick polyethylene.

BUTCHER
WRAP

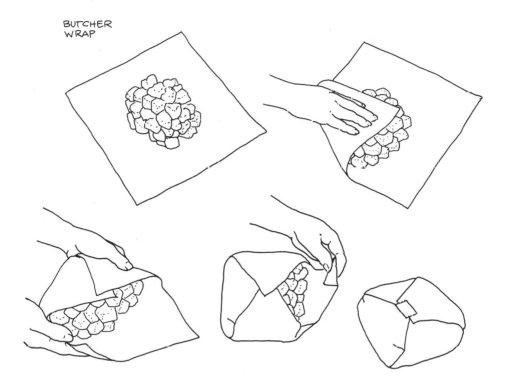

To make a butcher wrap, hold a large square of wrap with one corner facing you and place the food in the center of the square. Fold the corner nearest you up and over the food. The wrap should cover the food. Next, fold over the side corners, pushing them tightly against the food. Finally, fold down the corner opposite you, pressing it tightly against the food to remove air, and seal with tape.

Regular plastic wraps and wax paper tear much more readily and cannot prevent moisture loss. In fact, no product whose label does not specifically say that it can be used in the freezer should find itself around your food when you are squirreling things away for the long term. Don't use butcher paper, newspaper or cellophane. Don't use the packaging that meat and chicken are sold in, either—it's not intended for long-term freezer storage.

No matter how good a wrap you use, it won't be effective if you don't secure it properly around the food and seal it so as to exclude any chance that it could come unstuck. Aluminum foil does not need taping, because it forms its own seal if you mold it and fold it carefully, but freezer paper and plastic wrap do. This is not an area where creativity is encouraged. Elastic bands, Christmas tape, purple hair ribbons or the tape that helps you grip your tennis racket will not be happy in the freezer. Neither will glue, tent-seam sealer or thread. Use freezer tape, which is designed to go on sticking at below-zero temperatures.

Bags

Use freezer bags to collect assorted small cubes of frozen puree or herbs, or to hold separate scraps of frozen leftovers destined for the stock pot. Slide chickens, whole or in parts, or other oddly shaped foods, into bags. Freezer bags that close with a twist-tie are cheapest. (You can close them with soft twine, too.) They come in 1-pint, 1½-pint, 1-quart, 2-quart, 1-gallon and 2-gallon sizes. To keep a freezer bag from flopping around as you fill it, first slip it into the kind of holder that sometimes comes with kitchen scrap bags, or into a clean bucket or plant pot that can support the sides.

Once you've filled a freezer bag, you must then remove as much air as possible in order to prevent oxidation, which makes color, flavor and nutrients deteriorate. Garden-variety twist-tie bags leave it up to you to figure out how to do this. You can press the air out with your hand or suck it out with a straw. Or you can place the bag, filled but still open, into a bucket or other container that holds cold water. The water will force the air out of the bag before you seal it.

Once sealed, arrange the packages into rough squares for more compact storage in the freezer. Shape them with your hands, freeze them on cookie sheets, then stack them like bricks. Or you can slip a freezer bag into a freezer container, fill the bag, seal it, then place the bag, still in its container, in the freezer. Once the contents have frozen, remove the container, leaving a frozen block of food still in its freezer bag.

For about twice the price of freezer bags you twist-tie shut yourself, you can buy freezer bags that zip shut and form a very tight seal. These bags can be

When packing your food in plastic bags, remove all excess air. Place the filled bag into a bucket or deep bowl of water. Press the bag down into the water until all of the air has been expelled, then twist the top of the bag and tie shut.

reused, because they are made of a heavier material than standard freezer bags, and therefore are far less likely to break. These bags form thin packages that freeze quickly, thaw quickly and stack easily.

Further up the price scale of bags with special features are boilable bags that allow you to freeze blanched foods, thaw and reheat them all in the pouch. These bags are a lamination of polyester for boiling strength and polyethylene for sealing. They can make quick work of several food preparation processes (see Chapter 9), but they cost about twice as much as zip-closing bags and about four times as much as twist-tie bags. You can use boilable bags for foods with sauces to eliminate the need to scrape food off the bottom of the pot when you reheat, or to wash the pot at all. If you camp, you can pack food frozen in boilable bags in your food chest. It will help keep the rest of your food cold, and need only be reheated to make a quick meal.

Boilable-bag systems that are heat-sealed by machine have some other disadvantages, besides their price. The bags cannot be reused for freezing. You don't need the vacuum-seal feature available on some sealers unless you plan to keep food in the freezer for over a year. Besides, it must be used with extreme care because it does present a risk. Vacuum sealing creates an atmosphere free of oxygen. This anaerobic environment supports the growth of *Clostridium botulinum*, which produces the toxin botulism. If you leave a low-acid food on the counter in a vacuum bag before freezing it or while thawing it, the toxin-producing organism can grow. If you do use a vacuum-sealing machine, make sure the food in the bag is very clean, because *Clostridium botulinum* is a soil organism. Freeze the contents immediately, and open the bag to allow air in before you start thawing it or heat the contents directly from the freezer. If you use a water-drawn vacuum system, which creates a superior vacuum, the contents of the bag will cook more evenly because the bags will not float.

There are bags, and then there are bags. Not all plastic bags are suitable for freezing food. Don't freeze food in sandwich bags, bread bags, supermarket produce bags, wastebasket liners or garbage bags. Even the ones that might be safe for food will be far from strong enough to stave off the penetration of air and moisture.

Containers

Because a surprising number of foods have no particular shape of their own, we dollop them into rigid plastic containers, which do. In a perfect world, the plastic lids would not warp in the dishwasher or contract when frozen and lose their airtight seal. The containers would not crack, scratch easily or carry tomato sauce stains through life. But they do, and the most expensive brands can be as prone to mishap as the cheaper ones. Despite these problems, we all use containers for the kinds of foods that are soft at room temperature.

If you can, stock up on a quantity all of the same brand so you do not find yourself with a welter of tops and bottoms that do not fit each other. To use storage and freezer space efficiently, choose containers that nest when they are empty and that are square or rectangular, the optimum freezer shape, rather than round. Containers with twist-off lids, if you can find them, will eliminate your lid-shrinkage problems.

You can also use glass canning jars, but only the ones with straight sides. Food expands upward as it freezes and can crack jars with shoulders. The straight-sided ones also make it easy for you to overturn the jar and plop the food right out. Glass jars are cheaper than good plastic ones and sometimes they are free, if you use wide-mouth mayonnaise and peanut butter jars. Being glass, however, jars are usually round, harder to stack, and easier to break because they get slippery when they get cold. No freezer packaging is perfect.

Rigid containers are not confined to glass and plastic. You can also freeze your barley soup, lamb stew or chicken casserole in the pot you're going to warm it in later. Line the pot with heavy-duty foil. Cover the top with a separate sheet of foil and seal the edges all around. Freeze the lot, then lift out the contents and liberate the pot. Or skip the foil, freeze the contents in the pot, then transfer the solid block of food to a freezer bag. Put the food back in its pot to cook, without thawing it. Or do the freezing in an aluminum foil pan. Though you can't then heat the food in the microwave, the metal pan will do just fine in a conventional or convection oven.

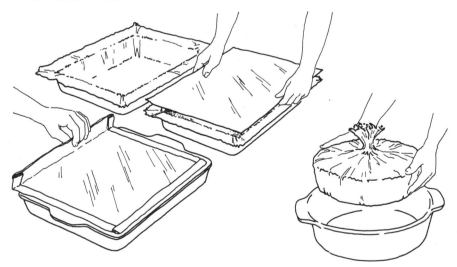

Gently press aluminum foil into a casserole or pot in which you plan to freeze the food. After cooking, use a separate sheet to cover the top of the dish, or simply gather the foil in the middle if using a round pot. Freeze the dish until firm, then lift the food out and return it to the freezer.

Freezing food in the cooking pot is a common practice for owners of microwave and convection ovens, who routinely gear their wrapping to how they will use the food. If you rely on a conventional oven, you can still use their example: Think ahead to how you will thaw the food. If you're freezing bread, wrap it in foil so you can thaw it in the oven right in its wrapper. Use freezer-to-oven cookware for foods that won't be staying in the freezer too long.

Think about quantities, too, when you package food for the freezer. Rather than having to take a hatchet to a hunk of pesto sauce the size of a brick, freeze it in small portions, wrap them in cheaper nonfreezer plastic wrap, then collect them all into a sturdy freezer bag or container. Use your baking containers to shape frozen food to manageable sizes. Freeze brown sauce in muffin tins, chili in ice cube trays, and soup in bread tins.

You can also freeze food in delicatessen containers, margarine tubs, cottage cheese or milk cartons, ice cream tubs, cans, cookie tins and coffee tins, *if* you line them with foil or plastic bags, or overwrap them with plastic bags. When using cans, line them so that the food does not touch the metal at all. When using a coffee can with its plastic lid, tape the lid all around with freezer tape to seal it.

Ice Cube Trays

Ice cube trays can be an especially convenient place to freeze certain foods. But using them regularly is a task for cooks who are particularly diligent about managing small amounts of food. If your cooking style is less measured, more spontaneous, freezing in ice cube trays may never become a habit because several other good habits are required. Nevertheless, here are some points to consider.

For one thing, you need a lot of ice cube trays. And once you've used the trays to freeze food, they won't work well for ice, because food odors will linger in the plastic. You also have to remember to transfer the frozen cubes into a bag; leave them in the trays and they'll evaporate in two or three weeks. Then, too, you have to remember to use them. Tomato paste is very handy in ice cube-size lumps but far less useful if you forget it's there for a year and a half.

You have to anticipate that cubes can easily get lost in the freezer, so you have to arrange not to lose them and not to grumble when you have to hike down to the basement to unearth them from the freezer. Finally, you must accept the fact that if the freezer temperature rises substantially, your cubes will quickly become soup on their own.

All that aside, if you can muster all these good habits—and not all freezer users can—you can reap the following rewards. As fast as others drop ice cubes into lemonade, you will be dropping your own tomato puree cubes into stew or gravy. With the speed of an ice cube hitting a glass of peach nectar, your homemade soup cubes of carrot and celery bits will hit the soup pot. Herb cubes you've made of fresh herbs pureed in the blender with a bit of water added will go splashing into your sauces or the water you cook grains in. Your baby will feast on baby-size portions of good food, vegetable purees and the like, from a source you can trust—your own kitchen.

The final verdict on freezing food in ice cube trays is that it means extra work for some cooks, extra convenience for others. The choice is yours.

A Few Words about Headspace

This book will often remind you to do two things that may seem contradictory. On the one hand it will caution you to wrap food tightly, excluding as much air as possible from the package to avoid having the air dry out the contents. On the other hand it will caution you to leave airspace at the top of containers. How come?

The reason is related to the contents. The airspace is a necessity in containers of liquid food, and solid food that includes juice, gravy or sauce. It allows space for the water contained in the food to move as it expands. Water expands 9 percent as it freezes, and it's a powerful, unstoppable kind of expansion. Water will shatter rock if it freezes in the confinement of a mountain crevice. High mountain peaks are normally rubble, not smooth stone, because water freezing in cracks has literally smashed the rock apart.

The force of water as it crystallizes into ice can exert 30,000 pounds of pressure per square inch if it is confined. If rock can't take this kind of pressure, imagine how your freezer containers will react if you pack them to the brim with food. Plastic containers crack, lids lift up and off, glass jars shatter, and your exposed foods dries out and winds up with slivers of glass or plastic embedded in it. Overfilled containers lead to a mess in the freezer.

The residents of Peking have no freezers. When the Chinese cabbage crop is harvested in December, the people stockpile hundreds of pounds of it at a time because it's the only green vegetable available all winter. They pile it up on balconies, on rooftops, under their beds, in their hallways and kitchens. They even hang it from their clotheslines!

To avoid problems, always leave headspace—½ inch in plastic pint containers and 1 inch in quart containers. If you use glass, which doesn't give the way plastic does, leave 1 inch in pints and 1½ inches in quarts. For easiest removal of frozen contents, use wide-mouth canning jars, jelly jars or peanut butter jars rather than jars with shoulders. If you do use jars with shoulders, measure headspace below where the jar narrows toward the lid.

The volume of solid food, unlike food containing liquids, doesn't change in the freezer, and any airspace you leave when you package it will only dry out your food. So, wrap chicken or mackerel or any other oddly shaped solid food in freezer paper, plastic or foil that will make it possible to press out all the air.

Labels

It's not enough to get your food into the right container. If you ever want to find it again, you have to label it. Be fair to your food. Don't resolve to freeze it today and definitely, without fail, label it tomorrow. Even very smart people immediately forget what's in the package they wantonly pitch, unidentified, into the freezer. When the miracle of freezing turns smooth applesauce to a solid block, you'll never figure out what it is without a label. Also, it's no fair changing the contents of a container and leaving the old label on. That will only lead to an instant identity crisis for the food and a big surprise for you when you defrost it.

Before you put anything in the freezer, take white freezer tape (it sticks better than freezer labels do) and write the following information on each package and container of food you freeze: what's inside, when it got there and how many people it will feed. That's so you can use the oldest food first, and food that is several years old you will know to dispense with altogether. If you are freezing many small items in a large bag, pop a slip of label paper inside the bag with the information facing out, of course. Always put the label where you can see it: on the top of the package if it's going in a chest freezer, on the side if it's going in an upright.

The good packaging that completely conceals your food in the freezer also stifles its chance to tempt you. Since the idea is to use what's in your freezer, and since once the food is there you can't see it, touch it or smell it, why not write yourself some mouth-watering messages about its contents?

If it's not a day when you're doing a blitz of freezer cooking, channel some of your energy into this creative labeling. You don't have to just put down the essentials and stop there. You can offer yourself suggestions on the best use of the ground beef you are packaging, to save yourself some time on a busy day when you want to use it. Or record in which stream you caught the trout before you. It's your label, and you can say what you please on it. Just say it with a felt-tipped permanent marking pen, not a pencil.

How to Keep Track ·
of the Freezer Inventory

You should be able to locate the pumpkin puree in your freezer with the same speed that you pluck a book from the shelf or a sweater from the drawer. But since you are dealing with a shifting inventory in the case of food, you're going to have to take your usual organizing habits one step further and keep written records. You're busy, so keep it simple. If your system isn't clear and quick to implement, you're likely to let it lapse.

Keep records of what goes in and what comes out, with a grease pencil on the freezer door. Or write it on a blackboard mounted nearby, or in a notebook

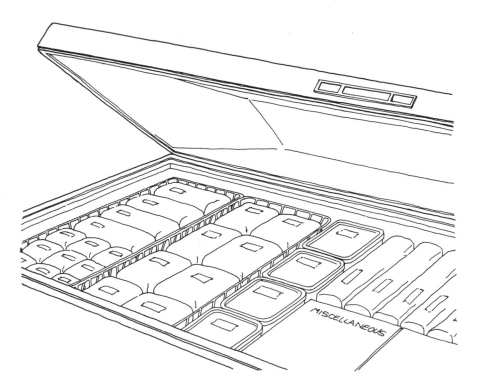

Keep an easy and accurate inventory of your freezer's contents by dividing the freezer into zones for each type of food and accurately labeling the packages.

When putting food in the freezer, tuck new packages under old ones with the same contents. That way, you'll be sure to use the older ones first. The sequence is always "first in, first out."

hanging on a hook. (Notebooks are good because they provide a permanent record for planning next year's garden and food purchases.) Or hang a list on the freezer, or make or buy a freezer map and fill in the spaces allotted for each food category. Add strawberry jam to the record as you drop it in, erase or cross out the stuffed mushrooms when you remove them for the party.

Think this kind of detail work is not for you? If you are trying to get along without it, you will discover that experience of chaos is the best teacher of order. If you decide to start an inventory system belatedly, do it when you defrost.

To keep track of where your food is in the freezer, divide your freezer into sections and assign a category of foods to each section, then stick with that order forever. If you put the lasagna on the bottom left every year without fail, you will always know where it is. (Don't assume you can put a mixture of foods on top so you can get at everything quickly. You'll just get to chaos more quickly.) For this zone system to work in a chest freezer, you will have to get yourself some baskets, boxes or clear plastic shoe boxes to act as dividers. Fill one box with your small, elusive packages of herbs, and another with your large, elusive packages of bread, rolls and muffins. Label another box "miscellaneous" and keep the odd egg white, ounce of lemon juice, dab of tomato paste or tablespoon of breadcrumbs in it until it adds up to something. In an upright freezer, you can use color coding. Use colored tapes or ties to indicate a category of foods, then heap them all together in one spot. If a corresponding color-coded map to your freezer contents hangs somewhere nearby, you're all set. To protect your hands while you're rooting around, wear oven mitts. They're as useful for handling cold foods as they are for hot ones.

How to Buy
Commercially Frozen Foods

It's not just frozen peas anymore. There are now thousands of frozen foods on the market, and there's something for everyone. A third of the new products are ethnic specialties—blinis, bagels, phyllo dough. It's a rare freezer owner who doesn't buy at least a few commercially frozen foods. It is freezer owners, after all, who have created this booming market, and the principal reason is that frozen food, in terms of flavor and nutritional content, is better than canned food. Just as you are likely to freeze a greater variety of foods at home than you're likely to can, you will also find a greater variety of frozen foods at the market than you do canned goods. If you buy frozen food, be alert to the freezer case you choose it from, the packaging that covers it and the ingredients that comprise it.

Let's get down to freezer cases first. The frozen food phenomenon creates some problems for supermarket retailers. Anywhere between 20 and 40 percent of their daily sales now come from refrigerated and frozen foods. It takes a lot of cases to house that food at the right temperatures, and the necessary equipment can cost as much as 70,000 dollars per store.

The store manager buying freezer equipment faces a trade-off between greater energy efficiency and lower visibility for the food. Chest cases display food poorly but are the cheapest to run, because the cold air stays near the bottom rather than rushing out to swap places with warm air in the store and drive up winter heating bills in cold climates. Upright freezers without doors make food very visible and accessible, but they cost the most to run. Store managers are increasingly choosing upright freezers with clear glass doors. They save 25 to 50 percent on energy costs, and that's the deciding factor.

Exact temperature control is also crucial. If a freezer runs at the wrong temperature for one day, it can thaw the contents enough to wipe out 3,000 to 4,000 dollars worth of food inventory.

While this kind of defrosting is obvious to everybody, other kinds are not so readily apparent. A store freezer may run slightly above 0°F or fluctuate occasionally in temperature, alternately thawing and refreezing its contents so slightly that the consumer cannot detect the process. It's also possible that somewhere between the plant, the warehouse and the store, the food may have thawed and refrozen.

How can you tell? If there's a thermometer visible in the case, first check to see that it reads no higher than 0°F. Then examine the packaging for clues. If the food is in a package that is soft or misshapen or has frost on it, don't buy it. The frost is a particularly good clue that the package has thawed enough at some point to allow moisture to escape from the food inside and freeze up on the outside. If the frozen food is in bags and you can feel clumps inside where there should be individual pieces, don't buy it. The clumps suggest that the contents have thawed and refrozen randomly in chunks. If you see pink ice or frozen juice in a package of frozen poultry, it means the contents have thawed and refrozen. Also, don't buy any frozen food in packaging that is torn or dirty.

Finally, examine the quality of the food itself before you buy it. Much of it contains salt, sugar and preservatives. To avoid undesirable ingredients, buy frozen foods that are simple rather than those that are breaded, in sauces or other fancy preparations. If you can find private labels (locally produced frozen food), it may be closer in terms of ingredients to what you would make at home.

Nutritionally, commercially frozen produce can actually be superior to fresh. Fresh supermarket produce may well come to you via slow truck, warehouse storage and long tenure in the display case. Its texture may be firmer than the frozen equivalent, but its nutrient value will have deteriorated along its slow route to your kitchen.

Producers of frozen produce, on the other hand, harvest crops at peak ripeness and process within hours at freezing plants located near the growing areas. They quick-freeze the food either in a cold solution or with blasts of frigid air that are much colder and therefore take effect much faster than the air in your home freezer. Such rapid freezing preserves quality well, giving ice crystals far less chance to form and nutrients almost no chance to dissipate.

On the other hand, frozen vegetables can sometimes contain a surprising amount of salt because of the way they are processed. For instance, when a solution of sodium chloride is used as the medium for sorting starchy vegetables like peas (the solution makes the big peas float and the little peas sink), the peas will retain some salt. Or when sodium hydroxide is used to chemically dissolve the skins of vegetables that require peeling, those vegetables will retain some salt. In these cases, salt will not even be listed as an ingredient on the package. It's all the more reason to grow and freeze vegetables yourself if you want to be very sure of what you are eating.

Along with commercial freezing processes, the addition of sauces to frozen vegetables can enormously enhance their sodium content. One-half cup of cooked cauliflower contains 6 milligrams (mg) of sodium. But over 3 ounces of commercially frozen cauliflower with cheese sauce contains 419 mg of sodium. Avoid particularly both frozen cauliflower and frozen broccoli when they are in cheese sauces, and frozen peas and brussels sprouts when they are in cream sauces.

Save the work of sorting your way through the frozen food aisle until the rest of your shopping is done, in order to minimize thawing. Once you've chosen and paid for your selection, put it in insulated bags, then collect all the bags into one big brown bag and hasten it home to the freezer. Your commercially frozen food will last a bit longer in the freezer than your homemade food. The quick freezing gives it a longer life and, unfortunately, so do its preservatives.

PART II

FREEZER
FOODS

CHAPTER 3

SAUCES
AND
SEASONINGS

Knowing what to expect from your freezer can make all the difference between successful meals and unforeseen disaster. Knowing what behavior to expect from the elements you add to your freezer foods—the gravies, sauces, herbs and spices—is crucial.

Sauces

You need to thicken sauces in a special way if you want them to freeze successfully. If you use conventional thickeners made from wheat flour, potato flour or cornstarch to make white sauces or gravies, they will curdle in the freezer. It is far better to thicken soups and stews after they've thawed, but if you wish to freeze a sauce or thicken a soup before you freeze it, there is a way to do it. Fortunately, there is Mochiko rice flour. If you cook for the freezer, you ought to have it on hand. Mochiko rice flour is a Japanese product, and it has the useful characteristic of stretching. It makes sauces smooth, and its elasticity allows it to freeze without curdling. Look for this flour in Oriental food stores or in supermarkets that serve Indonesian or Japanese neighborhoods. (See the recipe for Mochiko Rice Flour White Sauce on page 60.)

Don't overlook your own frozen vegetable purees as thickeners for your sauces and soups—they work extremely well. Add them to the foods you intend to freeze, and they will not separate out. They're also healthier than sauces based on butter and cream and thickened with flour, and at the same time they intensify the flavor rather than masking it.

There are some forms of sauces you should not freeze because they cannot withstand the chemical and physical changes freezing imposes on them. Don't freeze emulsified egg-based sauces, including hollandaise sauce and mayonnaise. The oil freezes more slowly and thaws more quickly than the other ingredients and ruins the emulsion. You can add eggs to a white sauce made with Mochiko rice flour, however, and freeze it successfully. Otherwise, don't freeze sauces with a high proportion of milk or cheese, because they will curdle.

Most of the sauces in this chapter take a different tack. They are designed to be easy and flavorful and freshly made. They are intended to add zest to the foods—especially vegetables—you use from your freezer. These sauces are so simple to prepare that they do not themselves warrant freezing. Their function is to enhance your frozen vegetables when you cook them.

Herbs and Spices

Freezing herbs, though not the most picturesque or the most convenient way to store them, is actually the best way to keep their flavor and color closest to fresh. The aromatic essence of herbs and spices is in their oils, and these oils respond to freezing with the happy habit of thickening up. Since the strength and character of the oil determines the flavor of the herb or spice, locking it in by freezing is a great advantage.

The process of drying herbs and spices, on the other hand, removes flavor as it removes the moisture. That's why frozen herbs are not always just the equivalent of dried herbs. Very often, frozen herbs are better, particularly in the case of chervil, chives, garlic chives, dillweed, fennel greens, parsley and tarragon.

Although your herbs would be most accessible lined up in jars on a shelf near the stove, or on the countertop, or growing in pots on a windowsill, these locations don't always work as well as we'd like them to. Though we associate herbs and spices with bright sun and warm breezes, once they're harvested sun and air are their worst enemies. Dried herbs should not be anywhere near direct sunlight, or near the heat of the stove. As for growing herbs in your kitchen,

If you understand how sauces and seasonings respond to freezing, you can use them with much more success when you cook for the freezer.

it's easy enough to do. The problem, if you cook with herbs a lot, is to grow enough of them on the windowsill to get you through a whole winter in a cold climate. The freezer, then, becomes your best best for herb storage during the winter. The improvement in quality is worth the few extra steps from stove to storage place.

The changes that take place when you freeze herbs and spices have more to do with appearance and texture than with flavor. You can expect that leafy herbs will darken, toughen slightly and go limp. You won't be able to use your frozen herbs as garnishes, but these changes don't matter when you cook with them.

Frozen herbs are simple to substitute for fresh in your favorite recipes. Take them straight from the freezer and add them to food in the same quantity that you would if they were fresh. No thawing or adjusting of measurements is necessary if the herbs were not blanched before freezing. Dried herbs and blanched frozen herbs are much more concentrated in flavor than are fresh or frozen herbs. If a recipe calls for fresh or frozen herbs, use only half as much of the same herbs dried. Herbs that were blanched before freezing should be measured like dried herbs. If a recipe calls for dried herbs and yours are fresh or frozen, use double the quantity specified.

How Herbs and Spices
Behave in Frozen Dishes

What happens to seasonings in the cooked foods that you freeze? The oils in the herbs and spices will slowly continue to permeate the food. Freezing won't stop them. The oils will bleed into the food until their essence has settled itself throughout. Since freezing gives the seasonings extra time to blend into the food, you can assume that the food you've frozen will emerge tasting different from the way it went in. Very often, the dish will taste more strongly seasoned when you serve it, not necessarily a bad thing, since many of us tend to underseason our food when we cook it.

It's the same principle you've seen at work when you've refrigerated a pot of chili for a day or two. It gets hotter and spicier, partly because the oils in the chili seasoning have had extra time to work. (By the way, there's another reason why food becomes more highly seasoned when you reheat it. The reheating invariably reduces the moisture content of the dish, thereby concentrating the

Make your own bouquets garni and use them straight from the freezer. The herbal essence that freezing locks in will disperse itself in a bubbling pot of stew.

flavors a bit more.) You may have read that some spices, black pepper in particular, should not be included in foods you intend to freeze. It is true that pepper intensifies in flavor in the freezer, but if you use it in moderation, it will not turn bitter and ruin the flavor of your food.

How to Store Spices in the Freezer

Since commercially ground spices lose their flavor and aroma quickly—within six months—many people like to grind their own. As long as spices are whole, you can safely keep them tightly capped in a kitchen cupboard. But when you release their oils by cracking or grinding them, you multiply their surface area at least eightfold and expose them to their enemy—oxygen. Then the spices belong in the freezer. (That's why ground coffee belongs in the freezer, too.)

If you're serious enough about the subject to keep a special coffee mill for spice grinding, and if you blend your own curry seasonings from whole spices, be sure to keep the leftover mixture and the partially used spices in the freezer. If your cooking style embraces one of the newest commercially available spices, green peppercorns, keep them in your freezer as well. Green peppercorns are highly perishable immature peppercorns. Even if you buy them packed in brine, you should expect green peppercorns to spoil. Either freeze them on a cookie sheet, collect them in a jar and keep them in the freezer, or add them to butter and freeze them.

There's one category of spices that belong in the refrigerator or the freezer for reasons of safety, as well as for flavor retention. Red spices, including red pepper, chili powder, all the paprikas and cayenne pepper contain insect eggs in their tissues, even after they've been ground up. When the temperature and the humidity are high, worms can grow.

Harvesting Herbs for the Freezer

Harvesting herbs for the freezer is fun. Wait for a time when you can work without interruption, harvesting and freezing in a single burst of sustained activity. Their flavor will last longer if your herbs go fresh-picked into the freezer. If you can, seize a moment early in the morning just after the dew has dried. Wet herbs are tricky to handle, and the more you inadvertently crush them during harvesting, the more surface area you expose to the air. If the sun has not yet gotten to the herbs, better still.

Lucky you if you are harvesting herbs that only gardeners have access to because they are not generally available commercially. You will be able to enjoy the special flavors of lemon thyme or apple mint all winter long. You can hold your herbs, whether they are rare or commonplace, for six months or more in the freezer. Take your herbs into the kitchen and get to work. If you're sure that they are free of soil, sprays and insects, you don't need to wash them. Otherwise, run them under tap water and shake off the excess. Pat them dry between towels or spin them gently in a salad spinner.

Blanching Herbs

Whether or not herbs need to be blanched before freezing is a matter of some debate. The Rodale Test Kitchen staff wanted to make sure. They froze basil, chervil, coriander, lovage, chives, thyme and dillweed, both blanched and unblanched. Their procedure for blanching was simple and fast. It involved gripping several stalks of the herb in question with tongs and quickly swishing them in a skillet of boiling water. The herbs were then spread on a towel to air-cool. None of the blanched herbs were subjected to cold water after the boiling water, on the theory that another dose of water might dilute their flavor. When they had cooled and dried, the herbs were frozen.

Basil provided the most dramatic results. The unblanched basil was black and smelled odd after freezing. But the blanched basil was bright green, tender and sweetly aromatic.

While basil was decidedly superior blanched, results with the other herbs were not so decisive. Chervil was not substantially improved by blanching, nor was coriander, because both froze very well anyway. Lovage froze especially well without blanching. The blanched chives were sweeter, but the unblanched chives were more oniony and had a better texture. Since chives are supposed

Pick up three or four herb stems with tongs, swish them carefully in boiling water for about five seconds, then drain on paper towels before freezing.

to be oniony, it's better not to blanch them. Blanched thyme had a better color and a truer aroma than the unblanched thyme. The unblanched had a grassy aroma. But frozen thyme, blanched or unblanched, was only marginally better than dried thyme. If you do freeze thyme, we recommend freezing whole springs, rather than just the leaves, to make it easier to handle. Dillweed, too, was no better blanched than unblanched, but it did freeze well.

While the color of herbs always improves with blanching, that is not reason enough to blanch herbs, since the color is lost again in cooking. There is a distinct advantage to blanching herbs, however, if you intend to hold them in the freezer for longer than six months. In that case, blanching will make a decisive difference in their quality.

The Many Ways to Freeze Herbs

Some herbs are best frozen while still on the stalk. Dillweed, in particular, is easiest to handle if you gather a whole bunch and freeze it in a freezer bag or container. When it's frozen, snip off bits as if it were fresh. If you freeze sage, rosemary and thyme on the stalk, they are that much easier to toss in the cooking pot and retrieve later.

Another alternative is to strip the leaves from the stems, rinse them and dry them thoroughly. Lay them out on a cookie sheet and leave them overnight in the freezer to freeze individually so they will not stick together when collected in one container. Then pack the leaves tightly into a freezer bag or container. Use the leaves straight from the freezer.

If you don't want to bother tray-freezing the leaves, they won't stay loose in the freezer, but you can easily break off a portion. At the Rodale Test Kitchen, cooks arrange herb leaves in a log shape for freezing, wrap them in plastic freezer wrap, then break off a corner as needed. And they don't bother chopping herbs before freezing them, knowing it's easy to chop them when they're frozen.

If you'd rather do more work on freezing day, dicing the herbs well before you freeze them gives you added convenience at cooking time. Packed into small containers or freezer bags, they'll be available for you to scoop out as you need them. The same diced herbs can go into ice cube trays, with a little water or stock to cover. Our tests indicate that the flavor of herbs frozen this way is consistently good. This method pretty much confines their use to soups and stews, however. You certainly can't add the cubes to salads, unless you don't mind picking the leaves out of a puddle of water. What you can do is to freeze some parsley or chervil in ice cube trays and use the cubes along with stock to make a quick cup of soup in the blender.

While you're washing, packing and generally messing around with herbs in the kitchen, it's a good time to take the whole process one step further. If the herbs you have on hand lend themselves to combining, make some ready-to-use mixtures, otherwise known as bouquets garni, and freeze them.

There are plenty of variations in what constitutes a bouquet garni. The most popular combination is a bouquet composed of three or four sprigs of parsley, two sprigs of thyme and half a bay leaf, tied together with kitchen twine. The bouquets, packed into freezer bags, can come out one at a time, to be dropped straight into sauce, stock or soup during the last minutes of cooking.

You can do the same thing with chopped or dried herbs. Collect them in a piece of cheesecloth about 4 inches square, and tie it up tightly with kitchen twine. To make enough of the herb mix to fill a dozen bags, combine 4 tablespoons of chopped parsley, 2 tablespoons of chopped thyme and four crumbled bay leaves. You could also add a few tablespoons of chopped celery leaves to the mixture, or a little marjoram. Freeze the bouquets and use them straight from the freezer. These infusions make good food better.

You can also make herb blends that you add to dishes a pinch at a time in the last moments of cooking. They infuse the dish with flavor, but themselves stay close to fresh. These blends are called fines herbes. The traditional mix is equal portions of parsley, chives and chervil, and half as much tarragon. Since all these herbs freeze better than they dry, make up small batches, well washed, dried and chopped, and freeze.

Here are some more special seasoning blends to freeze. For fish, mix equal quantities of thyme, basil, sage, sweet marjoram and crushed fennel seeds. For

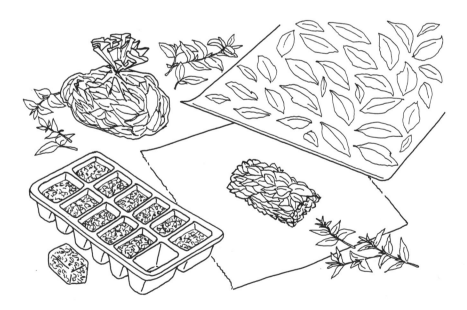

There are many ways to freeze herbs including, counterclockwise from lower left, ice cubes, log shape and individual leaves tray-frozen and then placed in plastic freezer bags.

poultry, game and meat, try equal quantities of sweet or French marjoram, basil, thyme and lemon thyme. For vegetables, mix sweet marjoram, basil, chervil and summer savory. For cheese, egg or potato dishes, soups and sauces, mix parsley, chives, chervil and tarragon.

As if those weren't enough ways to sort herbs for the freezer, there are still more. You can also freeze herbs in oil or in butter (see page 100). Unless you blanch it, basil in particular, loses its color in the freezer to a greater extent than most herbs, so to freeze it in oil is to give it a new home, in the form of Pesto (page 50). If pesto is the finest example of the oil-and-herb method, it is not the only one. You can add ¼ cup of oil to a cup of packed leaves of basil, or rosemary, sage, tarragon or thyme. Spoon the mixtures into ice cube trays. Once frozen, transfer them to freezer bags. Remember, when you add them to marinades and salad dressings, they are highly concentrated.

✳SAUCE DIABLO✳

A spicy sauce to dress up frozen fish, especially stronger-flavored, oily fish like bluefish. Can be frozen for 9 months.

¼ cup finely chopped onions	¼ cup red wine vinegar
1 hot chili pepper, split and seeded	¾ cup water
	2 teaspoons soy sauce
2 tablespoons olive oil	hot pepper sauce, optional
¼ cup tomato paste	

Sauté onions and chili pepper in oil for 5 minutes. Add tomato paste, vinegar, water and soy sauce and simmer, covered, for 20 minutes. If you like very hot dishes, leave chili pepper in the sauce. However, if the sauce is hot enough for your taste at this point, remove pepper. If the sauce is not hot enough, add a dash of hot pepper sauce.

Serve hot over fish, or freeze.

YIELD: 1½ cups

NOTE: Marinate fish fillets in the sauce for 30 minutes, turning occasionally. Remove fish from marinade, place on a broiler pan and broil about 8 inches from heat for about 5 minutes on each side, or until fish flakes when tested with a fork. Baste with sauce during broiling. Top fish with remaining sauce or serve sauce alongside as an accompaniment.

CHILI SAUCE

Use this very hot sauce as a condiment to spice egg dishes, stews and meats. Can be frozen for up to a year.

8 red chili peppers, split and seeded, thawed if frozen	4 cloves garlic
	2 cups chopped tomatoes, thawed if frozen
4 cups white wine vinegar, heated to boiling	4 small dried chili peppers

Combine red chili peppers, vinegar and garlic in a pint jar. Cover and let stand for at least 48 hours.

Combine the chili mixture with chopped tomatoes in a blender or food processor and puree.

Transfer to a double boiler. Add dried chili peppers and cook for 30 minutes over medium-high heat.

Remove dried chili peppers. Refrigerate or freeze.

YIELD: 1½ cups

MEXICAN CORIANDER SAUCE

Prepare this sauce for the freezer when the tomatoes are in season, to use as a topping for tacos, nachos and enchiladas. You can freeze it for 9 months.

2 cups fresh coriander	½ cup olive oil
3 cloves garlic	juice of 2 limes
1½ teaspoons dried oregano	6 large, fresh tomatoes, quartered
3 small hot chili peppers	
3 onions, quartered	1 tablespoon soy sauce

In the bowl of a food processor fitted with a steel blade, combine coriander, garlic, oregano and peppers. Using on/off turns, process until mixture is well minced.

Add onions, oil and lime juice; process with on/off turns until minced.

With machine running, add tomatoes. As soon as the last piece is put through the feed tube, turn off the machine.

Transfer mixture to a saucepan and bring to a boil over medium heat.

Add soy sauce and chill quickly by pouring into a bowl that is sitting in a sinkful of ice water. Stir occasionally, pack and freeze.

YIELD: 4 to 6 cups

∗SALSA VERDE∗

This green spicy sauce is often served at room temperature to top baked potatoes, rice or eggs. Tomatillos are available in Hispanic markets in the summer months. This sauce can be frozen for 6 months.

4 pounds tomatillos, halved	6 sprigs fresh parsley
½ cup water	4 sprigs fresh coriander or additional parsley
¼ cup corn oil	1 tablespoon soy sauce
1 medium-size onion	¼ cup cider vinegar
4 hot green chili peppers	1 teaspoon ground cumin
2 cloves garlic	

Simmer tomatillos with water and oil until limp and drab green. If very dry, add more water as needed.

Meanwhile, mince together onion, peppers, garlic, parsley and coriander, or process slightly in a food processor or blender.

Stir this mixture into the hot tomatillos, along with soy sauce, vinegar and cumin. Cool slightly.

Transfer to a food processor with a metal blade, or a blender, and process into a chunky sauce. Be careful not to make a smooth sauce. If using a blender process in four batches.

Chill for at least 4 hours before serving. Serve cool but not cold.

Freeze in small portions, as the sauce retains its fresh flavor for only 1 week after thawing.

YIELD: 6 to 8 cups

SALSA CRUDA

Salsa Cruda is a basic Mexican sauce that you can spoon in small amounts onto almost anything, including eggs, beans, tortillas and rice.

While this sauce is usually prepared with fresh tomatoes and left uncooked, it would separate if frozen this way. This version, tailored to freezer storage, is just slightly cooked and retains its fresh character. It's convenient, when working with chili peppers, to prepare a big batch at a time. The sauce can be frozen for 6 months.

6 cups finely chopped juicy fresh tomatoes (be sure to catch the juice)
½–¾ cup very finely chopped, seeded green chili peppers, about 8 long chilis (know how hot your chilis are before beginning and use discretion if very hot)

¾ cup finely chopped onions
¾–1 cup chopped fresh coriander leaves (no stems)
¼ cup lemon juice
1 tablespoon soy sauce
1 tablespoon honey
¼ cup Mochiko rice flour
2 cups tomato juice

In a large bowl, combine tomatoes, chilis, onions, coriander, lemon juice, soy sauce and honey. Stir well and let stand for 15 minutes for the liquid to come out of the tomatoes.

Meanwhile, place rice flour in a medium-size saucepan. Add tomato juice slowly, while stirring with a fork, to prevent the flour from lumping.

Press down lightly on the tomatoes to extract as much liquid as possible without bruising the tomatoes, and strain that liquid into the saucepan with the tomato juice mixture.

Bring to a simmer, stirring occasionally, over medium heat. When bubbly and thick, stir this mixture thoroughly into the tomatoes.

Pack into freezer containers, cool and freeze.

YIELD: about 4 pints

VARIATION: Substitute chopped fresh parsley for all or part of the coriander.

CARAWAY HONEY SAUCE

This is a fine sauce to toss with cubed or julienne cooked frozen carrots, beets or turnips, or steamed shredded cabbage. Do not freeze this sauce.

2 teaspoons caraway seeds, lightly crushed
2 tablespoons butter
2 tablespoons vegetable oil

¼ cup honey
3 tablespoons lemon juice

Over low heat, in a 1-quart saucepan, heat caraway seeds in butter and oil until seeds sizzle quietly, about 5 minutes. Add honey and lemon juice and stir well.

YIELD: ⅔ cup

FENNEL SEED AND ONION SAUCE

A good sauce for cooked frozen vegetables. Do not freeze the sauce.

1 large Spanish onion, coarsely chopped, about 1 cup
¼ cup butter
½ teaspoon fennel seeds, crushed in a coffee grinder or mortar with a pestle

⅛ teaspoon ground cinnamon
2 teaspoons lemon juice, thawed if frozen
1 tablespoon finely chopped fresh parsley

Sauté onions in butter over medium-high heat until slightly browned. Add fennel seeds and cinnamon, cover and cook over low heat. When onions are very tender, in about 15 minutes, add lemon juice.

Serve sauce hot over hot vegetables, topped with parsley.

YIELD: about ½ cup

LEMON-BASIL SAUCE

Serve this sauce over Broccoli, Cheddar and Chicken Packets (page 126). Do not freeze.

1 egg	1½ cups olive oil
juice of ½ lemon	ground black pepper,
2 tablespoons chopped	to taste
fresh basil	

Puree egg, lemon juice and basil in a food processor or blender. Add oil in a slow stream, as for mayonnaise, until mixture thickens. It will be runny, however—not as thick as mayonnaise. Taste and adjust seasoning.
      Serve at room temperature.

YIELD: 1½ cups

PESTO

This basil, nut and cheese sauce has become a freezer staple for gardeners who raise herbs. Proportions can vary; some people prefer to use parsley in place of some of the basil.
      Traditionally, pesto is made by hand with a knife and a mortar and pestle. Many cooks now use a food processor or blender, which creates a much smoother texture. Either way, the resulting pesto has many uses, from the basic linguini with pesto to seasoning cheeses and vegetables.
      Freeze pesto, complete with garlic and cheese, or add the cheese just before serving. Can be frozen for 9 months.

6 cloves garlic	½ cup grated Parmesan
4 cups lightly packed fresh	cheese
basil	½ cup grated Romano
¾ cup pine nuts or coarsely	cheese
chopped walnuts	1½-2 cups olive oil

To make pesto by hand, chop garlic coarsely and set it aside. Chop basil until the leaves lie flat and are no longer springy. Top basil with garlic and nuts and chop together until moist and crumbly looking. Transfer ingredients to a mortar and pound to a thick paste with a pestle. Stir in the cheeses with a fork. Add oil slowly, while stirring, to incorporate it as it is added. Add sufficient oil to make the pesto the consistency of a thick soup.

To make pesto with food processor or blender, with the machine running, drop garlic cloves one by one into the blade (if using a blender it's best to make this sauce in several batches; if using a food processor, use the metal blade). Stop, scrape and process again briefly. Add basil leaves and process, using on/off turns, until finely chopped. Add nuts and cheeses and use on/off turns again until mixture is moist and finely chopped. With machine running, add oil by tablespoons until pesto is the consistency of thick soup.

To freeze and thaw pesto, freeze in small portions; ½ cup is enough for 4 servings of pasta. Thaw in refrigerator, allowing about 6 hours.

To use pesto with pasta, use homemade or the best purchased pastas; the flat, thin shapes seem best. For 4 portions, boil ½ to ¾ pound pasta. Drain, saving ½ cup of the cooking water. Place pasta in a warm bowl. Stir a few spoonfuls of cooking water into ½ cup of the pesto and begin to spoon pesto onto pasta, tossing between additions. Add peas, cubed potatoes and a pat of butter if you wish.

Pesto is also good tossed with zucchini, broccoli and green beans. Or, spread fried eggplant slices thickly with pesto, top with chopped tomatoes and mozzarella cheese and bake until cheese melts.

YIELD: 2 cups

ALMOND BUTTER SAUCE

Swirl almond butter sauce with homemade or store-bought vanilla ice cream, and refreeze for several hours for a beautiful and tasty dessert. A food processor purees the almonds quickly and achieves the thin consistency necessary for mixing and freezing with the ice cream.

2 cups whole unblanched almonds	2 teaspoons maple syrup
5 tablespoons almond or peanut oil	¼ teaspoon almond extract
	⅛ teaspoon vanilla extract

Mix almonds with 1 tablespoon oil, until uniformly coated. On a large baking sheet, roast almonds in a 350°F oven until nicely browned, about 10 minutes. Cool slightly, 10 to 15 minutes. Reserve ½ cup roasted almonds.

In a blender or food processor, process remaining 1½ cups almonds with remaining oil to a thin, soupy consistency; this takes 3 to 5 minutes. Stir in maple syrup and almond and vanilla extracts by hand. Remove butter to a separate bowl.

Coarsely chop reserved almonds and stir into butter.

YIELD: about 1½ cups

MAPLE GLAZE

This is an excellent sauce for small onions or carrots. For starchy vegetables, such as sweet potatoes or winter squash, double the glaze ingredients. Do not freeze the sauce.

⅓ cup maple syrup
¼ cup frozen apple juice
    concentrate

Combine maple syrup and apple juice concentrate in a heavy saucepan. Cook over low heat until the mixture thickens, 5 to 10 minutes. Add desired vegetables to the syrup and simmer over low heat until tender, 15 to 20 minutes. Stir to glaze evenly.

YIELD: about ½ cup

RUSSIAN SORREL SAUCE

This tangy sauce for fish is really a French recipe with a Russian flavor. Do not freeze this sauce.

2 hard-cooked eggs
1 egg yolk
1 tablespoon cider vinegar
2 teaspoons prepared
     mustard
½ cup olive oil

¼ cup sunflower oil
⅓ cup Sorrel Puree
     (page 229), thawed if
     frozen
3 tablespoons chopped
     fresh parsley

Puree hard-cooked eggs and egg yolk in a blender or food processor. Add vinegar and mustard and blend again. With machine running on low, add olive and sunflower oils drop by drop until mixture begins to thicken, then add by teaspoons. Add sorrel puree and parsley and blend again until well mixed.

Serve over baked or poached fish.

YIELD: 1½ cups

MUSTARD SAUCE

There are few flavors so well adapted to perking up a vegetable as mustard. Try this sauce with cooked frozen green or wax beans, cauliflower, asparagus or broccoli. Do not freeze this sauce.

2 tablespoons butter
2 tablespoons whole wheat
     pastry flour
1 cup milk

3 tablespoons prepared
     mustard
½ teaspoon Worcestershire
     sauce

Melt butter in a small saucepan over low heat. Blend in flour to make a smooth paste. Stir in milk gradually, blending well. Continue stirring over low heat until thickened and smooth. Stir in mustard and Worcestershire sauce.

Serve hot over hot, cooked vegetables.

YIELD: 1¼ cups

SESAME MUSHROOM SAUCE

A light sauce to add to cooked frozen green beans, kale or any of your favorite greens. Do not freeze this sauce.

2 tablespoons corn oil	½ cup sesame seeds, toasted
6 mushrooms, sliced	1 tablespoon soy sauce

Heat oil in a skillet. Add mushrooms and sauté until tender. Add sesame seeds and soy sauce and stir gently.
Serve immediately by tossing with heated vegetables.

YIELD: ¾ to 1 cup

NOTE: To toast seeds or nuts, heat them in a dry skillet over medium heat, stirring occasionally, until lightly browned.

MARINARA SAUCE

This sauce requires a food mill to puree and strain the vegetable stock base. It can be frozen for 9 months.

1 large green or red bell pepper, seeded and chopped	10 leaves fresh basil, chopped, or 1 teaspoon dried basil
2 carrots, chopped	1-2 bay leaves
2 stalks celery, chopped	1 teaspoon dried oregano
1 large red onion, chopped	¼ teaspoon dried marjoram
⅓ cup olive oil	2-4 tablespoons tomato paste, optional
2 cloves garlic, minced	¼ cup freshly grated Parmesan cheese, optional
2½ pounds whole Italian plum tomatoes, thawed if frozen	
½ cup water, optional	
2 tablespoons chopped fresh parsley	

In a stockpot, sauté peppers, carrots, celery and onions in oil over low heat until just tender, about 12 minutes. Add garlic and sauté for 2 minutes more.

Add tomatoes and water (water is unnecessary if using frozen tomatoes), and bring to a boil over medium heat. Simmer, uncovered, over low heat for 45 minutes, stirring occasionally. Run mixture through a food mill.

Return strained mixture to stockpot. Add parsley, basil, bay leaves, oregano and marjoram. Simmer, uncovered, until sauce reaches desired consistency, 20 to 30 minutes. Remove bay leaves. Frozen tomatoes will make the sauce appear pale, so you may wish to add tomato paste for additional color and thickening.

Freeze now or stir in Parmesan cheese, if desired, about 5 minutes before serving.

Frozen sauce can be thawed first or, if still frozen, reheated slowly with a bit of water in a covered saucepan.

YIELD: 4 cups

SAUCE PROVENÇAL

This homestyle sauce, made with tomatoes, onions and garlic, is delicious with cooked beans, cubed summer or winter squash, sautéed eggplant or steamed zucchini slices. It can be frozen for 9 months.

¼ cup finely chopped
   onions
2 cloves garlic, minced
1 tablespoon butter
1 tablespoon olive oil
1½ cups peeled, seeded
   and coarsely chopped
   tomatoes, thawed if
   frozen

2-3 tablespoons tomato
   paste, optional
2 tablespoons chopped
   fresh parsley

In a large heavy skillet, cook onions and garlic in butter and oil over medium heat until golden. Add tomatoes and bring to a gentle boil. If necessary, thicken with tomato paste.

Toss with hot vegetables and sprinkle parsley on top or freeze for later use.

YIELD: 2 cups

BASIC TOMATO SAUCE

Once preserved only by canning, tomato sauce is now a freezer staple. This sauce will freeze for 9 months.

20 fresh, ripe plum
       tomatoes, coarsely
       chopped, about 8 cups
¼ cup olive oil
4 cloves garlic, minced
2½ cups chopped onions
2 bay leaves
½ teaspoon celery seeds

1 teaspoon fresh thyme
       leaves, or ½ teaspoon
       dried thyme
6 sprigs fresh parsley, finely
       chopped
   oregano, to taste
   basil, to taste
   cayenne pepper, to taste

Puree tomatoes in a food processor or blender.

In a large saucepan, heat oil and sauté garlic and onions. Add pureed tomatoes, bay leaves, celery seeds and thyme. Cook over low heat, uncovered, for at least 45 minutes, until thick. The longer this sauce is cooked, the better.

Add parsley and cook for 2 to 3 minutes. Season to taste with oregano, basil and cayenne. Remove bay leaves.

Pack into containers, cool and freeze.

YIELD: 4 cups

ITALIAN TOMATO SAUCE

Can be frozen for 9 months.

¼ cup olive oil
½ cup chopped onions
3 cloves garlic, minced
4 cups cooked, pureed
       tomatoes, thawed if
       frozen
3 cups Basic Tomato Sauce
       (page 56), thawed if
       frozen

2 cups sliced mushrooms
       (½ pound)
¼ cup chopped fresh parsley
1½ teaspoons dried oregano
1 teaspoon dried thyme
1 teaspoon dried basil
½ teaspoon dried rosemary
1 bay leaf
1 cup water

Heat oil in a 5- or 6-quart saucepan over low heat. Add onions and garlic and sauté until tender, about 4 minutes.

Stir in tomatoes, tomato sauce, mushrooms, parsley, oregano, thyme, basil, rosemary, bay leaf and water. Simmer, uncovered, for 1 hour, stirring occasionally. Cover and simmer, stirring occasionally, until thick and fragrant, 30 to 45 minutes.

To freeze sauce, pack in freezer containers, cool completely and freeze.

To reheat frozen sauce, thaw sauce completely and cook, covered, over medium heat, stirring occasionally, until hot, 15 to 20 minutes. Or reheat frozen sauce over very low heat, stirring occasionally, until hot, about 45 minutes. Serve over hot pasta.

YIELD: 8 cups

VARIATION: You can add Italian Meatballs (page 341) to the sauce before the final 30 to 45 minutes of simmering.

TOMATO PASTE

Can be frozen for 1 year.

    5 pounds tomatoes,
          preferably Italian plum

Quarter tomatoes and puree in a blender. Strain, if desired, to remove seeds and skins. Simmer over low heat in a large Dutch oven or shallow pot, stirring frequently, until thickened, 2 to 3 hours.

Freeze in small containers or ice cube trays.

YIELD: about 1½ cups

NOTE: Yield will vary somewhat, depending on the fleshiness of your tomatoes and whether or not you leave the seeds in.

BASIC MEXICAN TOMATO SAUCE

This is an easy-to-make, spicy, all-purpose sauce for serving hot or cold. This recipe is large, because it is handy to prepare a large batch while the vegetables are in season, but it is also possible to make the sauce with frozen tomatoes. You can freeze it for 9 months.

1 cup chopped onions	6 cups peeled, chopped
¾ cup diced bell or frying	plum tomatoes with
peppers	liquid, about 30
¼ cup peanut oil	tomatoes, thawed if
6 cloves garlic, minced	frozen
½–¾ cup seeded, diced green	ground black pepper,
chili peppers, about 6	to taste
long chilis (decrease	¼ cup chopped fresh cori-
amount if your chili	ander or parsley
variety is really hot)	1 tablespoon soy sauce

In a loosely covered saucepan, over medium heat, cook onions and bell peppers in oil until almost tender, 8 to 10 minutes. Uncover, add garlic and chili peppers and cook for 1 minute more.

Stir in tomatoes and black pepper and simmer, stirring occasionally, until thick, 20 to 30 minutes. Remove from heat and stir in coriander and soy sauce.

Pack into freezer containers, cool and freeze.

YIELD: 3 pints

NOTE: If you make this sauce with frozen tomatoes, peel them while still solid, thaw until soft enough to chop, and be sure to use all the juice.

SPINACH AND RAISIN SAUCE

Spinach and pasta have a great affinity for each other. Raisins and spinach provide an especially good contrast of flavors and textures. Serve over pasta or rice.

Pepare this for the freezer with fresh spinach (and freeze it for 2 months) or make it very simply with frozen spinach (do not refreeze).

1 medium-size onion,
   chopped
2 cloves garlic, minced
1 tablespoon olive oil
8 mushrooms, cut into
   thick slices
8 leaves fresh basil,
   chopped, or ½
   teaspoon dried basil
¾ pound fresh spinach,
   coarsely chopped,
   about 3 loosely packed
   quarts, or 2 cups frozen
   spinach, thawed and
   well drained

¼ cup golden raisins
2 tablespoons butter
   ground black pepper,
      to taste
¾ cup shredded mild
   Swiss or mozzarella
   cheese

Over medium heat, in a pot large enough to hold the spinach, cook onions and garlic in oil. When the onions begin to soften, add mushrooms and basil. Cook for a few minutes, until mushrooms become limp.

Stir in spinach and raisins. Cover pot and lower heat. Simmer until spinach is tender, about 10 minutes for fresh spinach, 4 to 5 minutes for frozen. Toss in butter and pepper.

If serving right away, cover pasta or brown rice with sauce, sprinkle with cheese and serve.

If freezing, pack in container, cool and freeze. To serve, thaw, then heat slowly in a saucepan.

YIELD: 2½ cups

MOCHIKO RICE FLOUR
WHITE SAUCE (THIN)

This sauce can be frozen for 6 months.

1 tablespoon butter, at room temperature	2 cups milk
1 tablespoon Mochiko rice flour	

In a small bowl, mix butter and rice flour together until a paste is formed (known as *beurre manié*). Bring milk to a boil. With a wire whisk, scoop the *beurre manié* and whisk vigorously into milk.

Whisk continuously until the sauce boils; lower heat and continue whisking for 4 minutes.

Season to taste.

Store in a jar in the refrigerator for up to 1 week. Freeze any sauce you won't be using in that time period.

YIELD: 2 cups

VARIATIONS: For a medium white sauce, use 2 tablespoons butter, 2 tablespoons rice flour and 2 cups milk. Proceed as directed.

For a thick white sauce, use 3 tablespoons butter, 3 tablespoons rice flour and 2 cups milk. Proceed as directed.

EASY WHITE SAUCE

This is an easy and slightly different way to make *beurre manie,* which produces an equally creamy, pleasant sauce. It freezes well for 6 months.

Select ingredients for thin, medium or thick Mochiko Rice Flour White Sauce (above). Over low heat, melt butter in 1¾ cups milk. Make a smooth paste of rice flour and remaining milk. Add flour mixture slowly to heated mixture. Stir continuously until thickened, about 4 minutes.

YIELD: 2 cups

HORSERADISH-CREAM CHEESE SAUCE

This sauce is good on just about anything but does wonders for the winter vegetables like cauliflower, brussels sprouts, carrots, red beets and cabbage, as well as for meat and fish. It can be frozen for 4 months.

3 tablespoons finely
    chopped sweet onions
1 tablespoon butter
⅓ cup Chicken Stock (page
    73) or vegetable
    stock (page 84),
    thawed if frozen
4 ounces cream cheese, at
    room temperature

ground black pepper,
    to taste
2 tablespoons freshly
    grated horseradish
2 teaspoons chopped fresh
    dillweed, optional

Cook onions in butter over low heat in a small saucepan for 1 minute. Add stock, cream cheese and pepper. Stir frequently until cheese melts and sauce is smooth. Remove from heat and stir in horseradish and dillweed.

Serve very warm. (If sauce becomes thin, a few minutes' cooling will thicken it again.)

YIELD: 1 cup

YOGURT-HORSERADISH SAUCE

A zippy sauce that will liven up any frozen cooked fish. Do not freeze sauce.

1 cup yogurt
1 teaspoon grated
    horseradish

1 tablespoon chopped
    fresh dillweed

In a small bowl, combine yogurt, horseradish and dillweed. Serve with fish.

YIELD: 1 cup

CHAPTER 4

SOUPS

Soup is an ideal food to freeze. It can stay frozen, in most cases, for up to a year and emerge with its texture intact and its flavor altered only subtly, if at all. There are only a few ingredients that lose their firm texture if you add them to soup before you freeze it, and these include potatoes, pasta, green peas and lima beans. These foods are best added at the last minute, when you reheat the soup. Also save any milk, cream, flour-based thickening agents and, of course, garnishes until the last minute. If you want to thicken soup before you freeze it, use sweet rice flour, called Mochiko rice flour (see page 38).

If you're making a soup that develops its flavor slowly in the pot, you can freeze it any time during the simmering process. When you reheat it, cook it some more. With most soups, cooking times aren't crucial. You can cook soup a little before it's frozen, and a lot after, or vice versa. In order to save space in the freezer, you can also cut down on the liquid content when you make the soup, and add more liquid when you reheat it. (See the recipes for soup bases later in this chapter.)

Taste the soup as you reheat it. Freezing can diminish some flavors and intensify others. If the garlic and onions have faded away, add more. This is the time to play with the soup a little. Add some fresh or frozen herbs that you might not have had around when you originally made the soup. Add leftover grains. You can also add frozen salad greens, which come out of the freezer much too limp to go into salads but do fine in soups, where flavor matters most and texture is secondary. You can serve up the same soup time and again, but with variations that turn it into a different dish each time and give the folks around your table a good sampling of your freezer's bounty.

Creating Soup from the Freezer

If you have a well-stocked freezer, you have the makings of a multitude of interesting soups. For instance, you can make soups from your frozen pureed vegetables and fruits (see pages 202 and 262). Thaw pureed cucumbers, add milk, buttermilk or cream, and you've got summer soup to transform a winter day. Or thaw pureed melon, stir in chopped fruit and yogurt, and you have a cool hot-weather soup. If you freeze your purees with a complementary herb —zucchini with basil, cucumber with dillweed—the soup will profit from the extra infusion of flavor.

Soups made from pureed vegetables are a blend of the thawed puree and a suitable liquid—stock, milk or the water saved from steaming or blanching vegetables. If the puree is a starch vegetable like carrots, peas or beans, it will need no further thickening. If the puree is nonstarchy—mushrooms, perhaps, or onions or leafy greens—it will need thickening as well, to give the soup body and texture. The thickener can be as simple as a few tablespoons of pureed leftover beans, some raw or cooked rice, macaroni, raw or mashed potatoes, or even bread crumbs. To thicken a broth and give it a little shine at the same time, mix a little cornstarch with water and stir it into the pot.

If you decide to thicken the soup with flour for a smoother texture, combine the flour first with butter, in the proportion of 1 tablespoon of flour to 1 tablespoon of butter for every 2 cups of soup. Melt the butter and stir in the flour. Cook gently for several minutes in order to brown the flour and dissipate its floury taste. Then mix in a little of the hot soup. Add that mixture to the soup pot and let it all simmer until it thickens.

At the last moment before serving, you can enrich the soup by melting a dollop of butter on top, adding some cream or incorporating a mixture of egg yolks and cream. Use 1 tablespoon of cream to one egg yolk for every cup of soup. Mix yolk and cream together, gradually stir in a cup of the hot soup, then return the whole mixture to the soup pot. Without letting the soup boil, stir it until it thickens to a velvety smoothness.

Think of every puree in your freezer as the basis for an interesting soup.

While soups based on a single dominant ingredient are often delicate and interesting, soup making from the freezer also lends itself to the creation of hearty vegetable soups with a multitude of ingredients. Make it a habit to scoop up cooked leftovers and freeze them with improvisational soup making in mind. One really simple way to make freezer soup is to add frozen vegetables to stock in the blender or food processor and puree. Season to taste and heat.

Soups made from and for the freezer don't have to be pureed, of course. There are a multitude of rich, full-bodied soups and stews that can be put together to form the basis of hot, hearty cold-weather meals. Take whatever odds and ends you've got stashed in the freezer, all the good scraps of meat, chicken, carrots and cabbage and cauliflower, cut them into bite-size pieces and recycle them back into soup. Proportions aren't sacred in homemade soup. In fact, that's what the word gumbo means—it's slang for everything you throw in the pot. You can use the leftovers and odds and ends in your freezer to concoct your own gumbos and stews.

If there's an ideal place for stashing leftovers, it's in soup.
If there's an ideal place for stashing soup, it's in the freezer.

Making Stock

The trick of turning your odds and ends into soup is to start with a good stock base. Traditional wisdom handed down among cooks says that good stocks make good soups and sauces, and it's true. Stocks are created themselves from leftovers. There's the basis for stock in the turkey carcass or fish bones that have been waiting in your freezer until you could attend to them. There's a potential stock in the year-old vegetables lurking in your freezer late in spring when the new garden season is beginning. They're taking up space you'll want for the new harvest, and they've been passed over so long they're probably no longer full of personality anyway. But don't toss those vegetables onto the compost heap. Instead, turn them into stock.

You don't have to make huge quantities of stock, either. If your freezer is not full of stock fixings, you can simply boil the leftover meat bones from dinner, or the remains of the pork chops or the chicken carcass. Freeze the special liquid that they create. It won't be a lot, but it will be delicious.

Unlike soup, which cools and goes straight into the freezer, stock made with beef, veal, lamb or chicken bones should spend a night in the refrigerator first. When the stock is cold, it's easy to skim off any fat and then freeze the stock. You need to remove the fat before freezing, because it has a limited life in the freezer and will turn rancid. If it's chicken fat you're skimming off, refrigerate it (page 122) and use it to sauté vegetables, or to make sauces and pâtés.

A quart of frozen stock will take two days to thaw completely in the refrigerator. It will thaw in an hour and a half in a bowl of warm water, which you should change as it chills. The same quart will thaw in 20 minutes in a covered saucepan over medium-low heat. For quick access, freeze some stock in ice cube trays and add the frozen cubes to stir-fries and other dishes where a little broth is all you need.

How to Make Bouillon Cubes

If your freezer space is limited, you can use it more efficiently by making bouillon cubes instead of stock. Unlike the ones you buy in stores, which are very salty, homemade bouillon cubes are a distillation of good, fresh ingredients. The four bouillon recipes in this chapter were developed by the Rodale Test Kitchen especially for the freezer. Unlike the stocks, which can be prepared quickly and left to simmer unattended for a time, bouillon requires careful tending and timing. Ingredients are added in a timed sequence so each has the opportunity to develop its flavor but is not in the pot long enough for its flavor to cook out. As the ingredients give up their flavor to the liquid, they create a strongly flavored stock.

No matter what combination of ingredients you use, the ones in the recipes that follow or your own mix of ingredients, the method of making bouillon

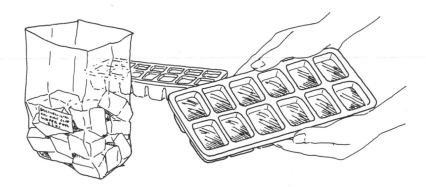

After the bouillon is frozen solid, transfer the cubes to plastic freezer bags. Secure the bag tightly and label it with the name of the bouillon and the date.

cubes is the same. If you are making the stock from meaty bones, include the skin and fat to add flavor to the liquid. The fat will be skimmed off later.

Cook the meat, vegetables and spices in water until all of the ingredients are very soft and have lost their flavor to the water. Then strain out the solids and chill the stock. Carefully remove the fat.

Cook down the stock at a rolling boil, until it appears thick and syrupy. If you prefer the bouillon to be clear, strain it through cheesecloth. Add the final seasonings, including delicate herbs and lemon juice, then cool. Pour into ice cube trays and freeze. When solid, transfer the cubes to plastic freezer bags. Label with the amount of water (usually about a cup) you will need to reconstitute the cubes. Once you have these convenient cubes on hand as a basis for soups and sauces, you will find that their versatility justifies the rather troublesome work of assembling all the ingredients to create them. One of the most pleasing ways to use your homemade bouillon cubes is to simply drop them back into water to make a hot soup to which you can add eggs, rice or pasta in the Italian style.

Techniques for Freezing and Reheating Soup

Always cool soup down before you freeze it. To hasten cooling and avoid the development of bacteria, set the soup pot into very cold water in the sink. Once the soup has cooled, pour it into freezer containers. Freeze soup in small quantities, 2 to 4 cups at a time, perhaps. It will freeze faster, thaw faster and eliminate the possibility that you will have to hack off a chunk with an ice pick to get a small quantity.

If you don't want to tie up your containers indefinitely, line them with freezer bags, pour in the soup, seal the bags and freeze. When the soup is frozen, release the containers that have acted as molds and stack the frozen bags of soup back in the freezer. If you are freezing soup in containers, leave an inch of headspace for expansion.

Frozen soup looks like a lump of ice, but don't let that deter you. It will thaw and reheat beautifully. To reheat frozen soup (and other foods high in liquid content), just run the container under warm water to quickly loosen the contents, then plop it straight into a saucepan. Add ½ inch of water to the bottom of the pan to hasten thawing (by creating steam) and to help prevent scorching. Keep the heat low and stir the thawing soup gently to further prevent sticking. Keep a lid on to help the flavors blend well and prevent the liquid from evaporating. Don't let the soup boil, but make sure it is heated thoroughly before serving.

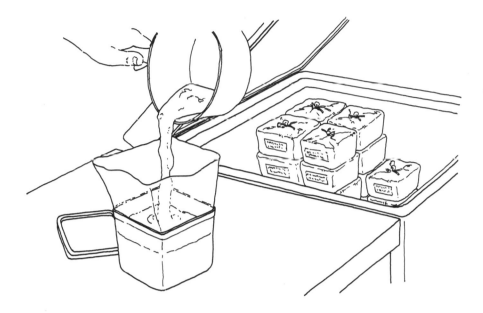

Place a plastic freezer bag into the size container in which you want to freeze your soup. Use a straight-sided container, which makes removal of the frozen soup easier. Carefully pour the soup into the bag, leaving enough room at the top for expansion. Secure the bag and place the container in the freezer. When the soup is frozen, remove the bag from the container. Label the soup, note the date and stack like bricks to save space in the freezer.

FROZEN GLACÉ

In classical French cooking, highly reduced stocks are used as flavoring agents for sauces, to enrich or extend soups and for many other cooking purposes. Glacé can be reconstituted to stock by adding water to it, or can be used as is for an intense flavor. It can be frozen for 9 months.

2 quarts stock, thawed if
frozen

Reduce the stock to 1 cup of liquid by boiling it down. As the quantity of liquid is reduced, pour it into a smaller pot so it will not burn. The reduction will take about 1 hour. Freeze the resulting thick, dark syrup in ice cube trays and, when frozen, store in plastic bags.

YIELD: 6 to 8 ice cubes

MONASTERY BEAN SOUP

Bulgarian monks used to prepare this garlicky soup to feed faithful pilgrims. Now they offer it to the tourists who come to visit their isolated mountain retreats. The monks are outstandingly vigorous, and many are over 100 years old. Their soup is thinner than most bean soups, and it is served with a splash of a red wine vinegar and a pat of butter. The soup can be frozen for 9 months.

¾ cup dried navy beans, soaked in hot water for 2 hours or in cold water overnight (or see (Beans, Dried, page 204)

6 cups beef broth or pork broth, thawed if frozen, or water

1 cup coarsely chopped onions

1½ cups diced carrots

2 tablespoons butter

¾ cup seeded and diced yellow or red frying peppers, 2 to 3 peppers

5 cloves garlic, finely chopped (but not minced)

½ teaspoon caraway seeds

½ cup chopped tomatoes, thawed if frozen

1 tablespoon soy sauce butter

2 tablespoons finely chopped fresh parsley croutons red wine vinegar

Drain beans and simmer in broth in a medium-size saucepan for 25 minutes. If you are preparing this soup for the freezer, you can cut down on the liquid and add it when you reheat.

Meanwhile, sauté onions and carrots in butter. When onions are lightly browned, stir in peppers, garlic and caraway seeds. Cook for 1 minute more.

Add a ladleful of the bean liquid to the onions. Deglaze the pan by stirring well to loosen the brown bits from the sides. Then add onion mixture to beans. Simmer, covered, until beans are tender, 20 minutes or more. Thicken the soup a little by crushing some of the beans with a wooden spoon or potato masher. Add tomatoes and soy sauce and cook for 5 minutes more.

If you want a pat of butter in your soup, put it on now.

Sprinkle each serving with parsley and croutons. Pass vinegar at the table.

YIELD: 6 to 8 servings, about 8 cups

MINESTRONE

Minestrone fits beautifully into a freezer routine. You can prepare the entire soup during the gardening season and freeze it, omitting the water to save space. Or, assemble it from frozen ingredients during the winter and refreeze it. It can be frozen for 9 months.

1 cup chopped onions	1½ cups cooked white beans
1 cup chopped green	or garbanzos
peppers	½ cup uncooked, small,
3 cloves garlic, minced	whole wheat pasta
3 tablespoons olive oil	1 cup chopped tomatoes,
1 cup chopped celery	including juice,
1 cup chopped carrots	thawed if frozen
1 cup chopped zucchini	4 cups water
1 tablespoon kelp powder	¼ cup chopped fresh
or vegetable seasoning	parsley
dash of cayenne pepper	Parmesan cheese, or ¼
1 teaspoon dried oregano	cup Pesto (page 50),
2 teaspoons fresh basil, or 1	thawed if frozen
teaspoon dried basil	

In a large heavy pot, sauté onions, green peppers and garlic in oil until wilted. Add celery, carrots, zucchini, kelp powder, cayenne, oregano and basil and mix well. Cover and let steam for 10 minutes.

Add beans, pasta, tomatoes and (if serving right away) water. Bring to a boil and simmer until vegetables are just tender.

At serving time, sprinkle with parsley and pass the Parmesan cheese or place a spoonful of pesto in each bowl before filling with soup.

YIELD: 8 to 10 servings, 10 cups

BEEF BOUILLON CUBES

Use fresh or frozen vegetables and meats for this bouillon. You can freeze it for 12 months.

4½ pounds meaty shin beef, with bone, sawn into 1½-inch slices, thawed if frozen
1 pound veal knucklebone
3 cups sliced onions
3 tablespoons red wine vinegar
6 quarts water
8 medium-size carrots, cut into large pieces
3 cups coarsely chopped celery, including leaves

8 mushrooms, halved
½ teaspoon black peppercorns
8 bay leaves
1 tablespoon caraway seeds
1 small dried chili pepper
4 cloves garlic, halved
2 teaspoons fresh thyme leaves, or 1 teaspoon dried thyme
3 tablespoons soy sauce, optional

Place shin beef, veal knucklebone and onions in a single layer in a shallow roasting pan. Broil, turning occasionally to brown all sides. Place meat bones and onions in a stockpot.

Deglaze the roasting pan with vinegar, scraping the bottom well to loosen any browned bits. Add to the stockpot along with the water, carrots, celery and mushrooms.

Cover and simmer until beef is very tender and falls from bones, about 3 hours. Uncover and add peppercorns, bay leaves, caraway seeds and chili pepper. Simmer for 30 minutes and add garlic and thyme. Cook for another 30 minutes.

Cool slightly and strain through a colander into a large bowl. Add soy sauce.

Chill until fat congeals. Remove the fat. At this point you should have 6 quarts or more. Return liquid to heat and cook down, at a gentle boil, to 1½ quarts. If the stock has a lot of sediment at this point, strain it through a piece of wet cheesecloth.

Cool, pour into 3 ice cube trays and freeze. When frozen solid, transfer the cubes to plastic bags. One cube will flavor 1 cup of liquid.

YIELD: 42 cubes

BEEF STOCK

A rich, meaty-flavored stock that freezes well for 12 months.

6 pounds beef bones
2 cups 2-inch chunks
     onions
1 cup 2-inch chunks celery
1 cup 2 inch chunks carrots
9 quarts water
¾ cup chopped tomatoes,
     partially thawed if
     frozen

1 clove garlic
1 bay leaf
2 whole cloves
6 black peppercorns
1 teaspoon fresh thyme
     leaves
3 sprigs fresh parsley

In a roasting pan, broil beef bones and vegetables, stirring so that all sides are browned evenly. Transfer bones and vegetables to a stockpot, pour the fat out of the roasting pan and deglaze with 1 quart water. Add this water and remaining water to the stockpot, along with the tomatoes and seasonings. Bring to a boil, skimming off the foam occasionally.

Reduce heat and simmer for 4 to 6 hours, uncovered, skimming as often as possible. The more you carefully skim the stock, the clearer it will be.

Strain the stock through wet cheesecloth. Refrigerate. Skim off the fat when cool and freeze.

YIELD: about 6 quarts

CHICKEN BOUILLON CUBES

Use fresh or frozen vegetables. Chicken necks, backs and wings can substitute for a whole chicken. Can be frozen for 12 months.

6 pounds stewing chicken parts, thawed if frozen
3 large onions, cut into thick slices
2 pounds carrots, cut into 2-inch chunks
4 cups coarsely chopped celery (including the leaves from the center of the heart)
2 bell peppers, coarsely chopped
1 cup chopped tomatoes
8 quarts water
10 bay leaves
½ teaspoon black peppercorns

6 cloves garlic, halved
2 tablespoons fennel seeds
2 tablespoons caraway seeds
2 teaspoons celery seeds
2 tablespoons crumbled (not rubbed) dried sage
1½ teaspoons paprika
2 tablespoons chopped fresh basil, or 1 table-spoon dried basil
¼ cup chopped fresh parsley
1 teaspoon fresh thyme leaves
3 tablespoons lemon juice

To brown the chicken and the onions, place pieces in a single layer in a large roasting pan. Broil, turning occasionally, until all sides are well browned. A little charring is acceptable.

Put chicken and onions, along with carrots, celery, bell peppers and tomatoes, in a stockpot. Use some of the water to deglaze the roasting pan, scraping the pan carefully to dissolve all the drippings. Add to the stockpot along with remaining water.

Simmer, covered, until meat falls readily from the bones, about 3 hours. Uncover, and cook for 1 more hour. Strain stock through a colander into a bowl.

Let broth cool until fat congeals and can be removed. The broth should measure about 6 quarts.

Return broth to stockpot and add bay leaves and peppercorns. Cook, uncovered, at a slow boil until stock is reduced to 1½ quarts. The stock appears slightly syrupy at this point and is strongly flavored. Add garlic and fennel, caraway and celery seeds. Cover pot and simmer slowly for

20 minutes. Strain out spices. Then add sage, paprika, basil, parsley and thyme. Simmer for 10 minutes more.

Chill, add lemon juice, then pour into 3 ice cube trays and freeze. When frozen solid, transfer the cubes to plastic bags. One cube will flavor 1 cup of liquid.

YIELD: 42 cubes

CHICKEN STOCK

Can be frozen for 12 months.

8 pounds chicken bones, backs and necks, thawed if frozen	2 cloves garlic
	1 bay leaf
	3 whole cloves
2 cups coarsely chopped onions	12 black peppercorns
	1 teaspoon fresh thyme leaves
1 cup coarsely chopped celery	6 sprigs fresh parsley
1 cup coarsely chopped carrots	8 quarts water

Wash chicken parts. Place chicken, onions, celery, carrots, garlic, bay leaf, cloves, peppercorns, thyme and parsley in a stockpot and add water. Bring the stock to a boil. Skim foam from the surface. Reduce heat and simmer the stock, uncovered, for 4 hours, skimming occasionally. The more the stock is skimmed, the clearer it will be.

Strain through wet cheesecloth. Refrigerate, skim off the fat when cool and freeze.

YIELD: about 4 quarts

NOTE: To make a brown chicken stock, you can brown the chicken parts and vegetables, as for Beef Stock (page 71).

DILL CABBAGE SOUP

A quick and hearty soup, this one takes 15 minutes to prepare and 15 minutes to simmer. If you prepare the soup with fresh cabbage, you can freeze it for 9 months with all or part of the liquid. It is equally good prepared with frozen cabbage. If using cabbage frozen in wedges, partially thaw before chopping.

2   cups coarsely chopped cabbage, partially thawed if frozen
¾   cup coarsely chopped onions
1   teaspoon dill seeds
½   teaspoon caraway seeds
2   tablespoons butter
3   cloves garlic, minced
1   tablespoon red wine vinegar
1   tablespoon red or white grape juice

1½   cups tomato juice
2   teaspoons soy sauce
3   cups water
1   large potato, diced ground black pepper, to taste
6   tablespoons yogurt or sour cream
2   tablespoons finely chopped fresh parsley or dillweed

In a medium-size saucepan, sauté cabbage, onions, dill and caraway seeds in butter, stirring occasionally, until the cabbage is translucent and wilted, about 10 minutes.

Add garlic, vinegar and grape juice. Cook for 1 minute and add tomato juice, soy sauce, water and potatoes. Cover and simmer until potatoes are tender, 15 to 20 minutes. Add pepper.

Top each serving of soup with a spoonful of yogurt and sprinkle with parsley.

YIELD: 6 servings, 6 to 7 cups

ZUCCHINI SOUP BASE
FOR THE FREEZER

Soup is one of the best uses for zucchini because the delicately flavored liquid that drains from the zucchini flavors the soup well. This soup base can be frozen for 6 months.

4 cups sliced fresh zucchini
2 cups Chicken Stock (page 73), thawed if frozen
½ teaspoon freshly ground black pepper
2 teaspoons chopped fresh basil, or 1 teaspoon dried basil
1 tablespoon chopped fresh celery leaves, or ½ teaspoon dried celery leaves

1 cup chopped onions
⅓ cup butter
3 tablespoons whole wheat flour
3 tablespoons Mochiko rice flour (or use all whole wheat flour if you won't be freezing the soup)

Combine zucchini, stock, pepper, basil and celery leaves in a large saucepan. Cover and simmer until zucchini is tender.

Meanwhile, sauté onions in the butter over low heat in a skillet until slightly browned. Stir in wheat and rice flours. Cook for 1 minute more. Stir this into the zucchini mixture and cook until thickened, stirring constantly.

Cool and freeze in 2-cup containers.

YIELD: about 4 cups

SOUP USING FROZEN ZUCCHINI SOUP BASE

2 cups frozen Zucchini Soup Base (page 74), thawed
2 cups milk
½ teaspoon ground nutmeg

pinch of ground black pepper
½ cup shredded Swiss cheese

Combine zucchini base, milk, nutmeg and pepper in a medium-size saucepan. Cover loosely and cook over medium heat, stirring occasionally, until smooth and hot, but not boiling. Add cheese and stir until just melted.

YIELD: 4 to 5 servings, 4 cups

NOTE: The soup base can be thawed quickly by placing the container in warm water.

FISH CHOWDER BASE
FOR THE FREEZER

Milk-based chowders can be frozen after they are prepared but will keep only a few months. It is better to assemble a base for chowder, freeze it, and add the milk and potatoes at serving time. Another possibility is to freeze fish and fresh-shucked clams in separate packages and assemble the soup from these. This chowder base can be frozen for 6 months.

24 chowder clams	1 tablespoon fresh thyme
1 cup water	leaves, or 1 teaspoon
1 cup chopped onions	dried thyme
1¾ cups diced carrots	2 pounds cod fillets or a
1½ cups chopped bell	mixture of white fish
peppers, red and green	fillets, cut into 1-inch
mixed	chunks
⅓ cup butter	
2 bay leaves	
1 teaspoon chopped fresh	
tarragon, or ½ teaspoon	
dried tarragon	

Scrub clams well and place in a steamer with water. Steam until clams open, 12 to 15 minutes. Remove from heat; strain off and reserve liquid. Remove clams from their shells and finely chop. Set aside.

Sauté onions, carrots and peppers in butter in a covered saucepan for 10 minutes. Add bay leaves, tarragon, thyme and reserved clam broth. Simmer for 10 minutes. Add cod, cover and remove from the heat. Cool and add clams. Remove bay leaves. Pack in 1-pint portions and freeze.

YIELD: 8 cups, enough for 5 quarts of soup

CLAM CHOWDER USING
FROZEN FISH CHOWDER BASE

Here's a tasty New England–style clam chowder made with the frozen base in the preceding recipe. Do not freeze this chowder.

1   cup milk
2   tablespoons cornstarch
1   cup light cream
2   cups frozen Fish Chowder
       Base (page 76),
       thawed

2   cups unseasoned
       mashed potatoes
¼   teaspoon ground white
       pepper
1   tablespoon chopped
       fresh parsley

Stir 3 tablespoons of milk into cornstarch until smooth, and set aside.

Combine remaining milk, cream, chowder base and mashed potatoes in a large saucepan. Cook over medium heat, stirring occasionally, until hot. Stir in the cornstarch paste and cook until chowder is smooth and steaming hot. Add pepper and parsley and serve with crackers.

YIELD: 4 to 5 servings, 5 cups

NOTE: The quickest way to thaw the chowder base is to place the container in a bowl of warm water.

VARIATION: To prepare a Manhattan-style chowder, cook 1½ cups diced potatoes until tender in 1 cup of chopped tomatoes and 1 cup of tomato juice. Substitute this for the milk, cornstarch, cream and mashed potatoes. You may wish to add corn or peas as well.

FISH STOCK

While fish stock is not as commonly used as other stocks, it is neverthe-less essential in making fish soups and sauces. Fish stock can also be used to poach or bake fish and to make aspics to glaze whole fish that will be served cold. Since making fish stock is undeniably going to make the kitchen smell fishy, it is a good idea to make a large batch at one time. Accumulate only very fresh fish bones and heads, and shrimp shells, either raw or cooked, and freeze them until you have enough to make stock. Some stock can be left at the right strength for soup, and the remain-ing stock can be cooked down to make a more intense base for sauces.

You can use fresh or frozen vegetables in this stock, and the stock can be frozen for 12 months.

1 cup sliced carrots	1 tablespoon rice vinegar
1 cup chopped Spanish onions	3 pounds fish bones and heads (must be very fresh)
½ cup chopped red bell peppers	3 quarts water
½ cup chopped celery	2 sprigs fresh thyme leaves, or 1 teaspoon dried thyme
3 tablespoons butter	
3 tablespoons white grape juice	

Cook carrots, onions, peppers and celery in butter in a large pot, lightly covered, until they begin to brown. Add grape juice and vinegar. Stir well to loosen browned bits from the bottom.

Add fish bones and heads, water and thyme and simmer, covered, for about 1½ hours, until the bones and heads are softened. Strain and freeze at this point or cook stock, uncovered, until volume is reduced by half.

YIELD: about 2 quarts stock, about 1 quart reduced stock

HERB BOUILLON CUBES

This is an excellent way to preserve the flavor of fresh herbs. Use the cubes to season soups, stews and sauces. Try different herb combinations for special uses—for instance, an Italian herb combination with basil, rosemary, oregano and parsley. The bouillon can be frozen for 12 months.

3 cups Chicken Stock (page 73) or vegetable stock (page 84), thawed if frozen
1 bouquet garni (1 teaspoon cumin seeds, 1 tablespoon anise or fennel seeds, 1 teaspoon celery seeds, 2 bay leaves, chopped, 3 cloves garlic, slivered)
¼ cup tomato juice
2 teaspoons finely chopped fresh sage, or 1 teaspoon dried sage
2 teaspoons chopped fresh rosemary leaves
2 teaspoons fresh thyme leaves
½ teaspoon paprika
1 clove garlic, finely chopped
¼ cup finely chopped fresh basil
2 teaspoons dried oregano
1 teaspoon dried marjoram
1 teaspoon chopped fresh dillweed
3 tablespoons finely chopped fresh parsley
4 teaspoons finely chopped fresh chervil
2 tablespoons chopped fresh chives
1 tablespoon lemon juice
dash of cayenne pepper
dash of turmeric

In a large pot, simmer stock and bouquet garni for 15 minutes. Remove bouquet garni.

Add tomato juice, sage and rosemary and cook for 10 minutes. Add thyme, paprika and garlic and cook for 5 minutes. Cool slightly, then add basil, oregano, marjoram, dillweed, parsley, chervil, chives, lemon juice, cayenne and turmeric.

Pour mixture into ice cube trays, cover with wax paper and freeze. When frozen, transfer the cubes to freezer bags. One cube will flavor 1 cup of liquid.

YIELD: 28 to 36 cubes

ONION OR LEEK SOUP BASE
FOR THE FREEZER

While onions are not typical candidates for the freezer, sometimes freezing them is the best alternative. That's the case if you find it convenient to slice a large quantity at one time, particularly if you use a food processor. If your onions are bruised or improperly cured, freezing is also the best way to store them. Spanish onions are not good keepers, and freezing may be the only way to store an overabundance.

An ideal use for a quantity of onions is as a base for onion soup. Leeks can be used instead of onions for a more fragrant and delicate soup.

This recipe makes enough base to prepare 4 to 6 quarts of soup. The base can be frozen for 6 months.

5 pounds onions or leeks	4 teaspoons fresh thyme
2 tablespoons vegetable oil	leaves, or 2 teaspoons
½ cup butter	dried thyme
¼ cup whole wheat pastry	water
flour	2 tablespoons Dijon-style
2 tablespoons Mochiko	mustard
rice flour	3 tablespoons soy sauce
1 tablespoon chopped	
fresh rosemary leaves,	
or 2 teaspoons dried	
rosemary	

Cut onions into thick slices lengthwise and cut these into 1-inch lengths. If using leeks, remove roots and tough green leaves, quarter them lengthwise and chop into 1-inch lengths. Wash them thoroughly to remove sand.

Heat oil and butter in a large pot over medium heat. When foaming of the butter subsides, add onions, cover the pot loosely and cook, stirring occasionally, until they are lightly browned, about 25 minutes. Add wheat and rice flours, rosemary and thyme and, stirring occasionally, cook until the flour is also lightly browned, another 8 minutes. Add just enough water to cover onions and stir until smooth. Stir in mustard and soy sauce.

Pack in 1-cup containers, cool and freeze.

YIELD: 4 to 6 cups

SOUP USING FROZEN ONION OR LEEK SOUP BASE

Here's how to use the preceding frozen soup base to make a delicious finished onion or leek soup.

8 thin slices day-old Italian bread	3 cups water
softened butter, as needed	2 cups shredded Gruyère cheese
1 cup frozen Onion or Leek Soup Base (page 80), thawed	½ cup grated Parmesan cheese

Butter bread thinly, place on a baking sheet and bake in a 300°F oven until crisp, 15 to 20 minutes.

Meanwhile, in a large pot, bring soup base and water to a boil.

Place 1 slice of crisped bread in the bottom of each of four oven-proof soup crocks. Top with some of the cheese (both kinds). Fill each bowl with soup, leaving at least 1 inch at the top. Top with remaining bread and cheese. Bake in a 400°F oven until soup is bubbly hot and cheese is melted and slightly crusty.

YIELD: 4 servings

NOTE: The quickest way to thaw the soup base is to place the container in a bowl of warm water.

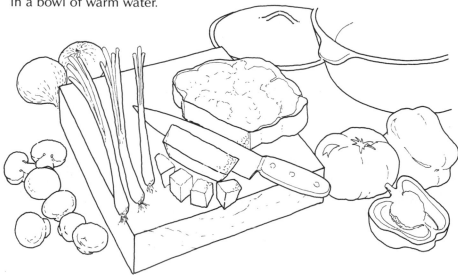

VEGETABLE BOUILLON CUBES

Any vegetable, fresh or frozen, can be used in this bouillon, but use discretion with stronger-flavored vegetables such as cauliflower, broccoli and turnips. The bouillon can be frozen for 12 months.

3½ cups coarsely chopped onions	2 cups diced potatoes
1 tablespoon vegetable oil	4 cups chopped leeks, including green leaves
4 cups coarsely chopped carrots	10 bay leaves
5 cups coarsely chopped celery, with leaves	2 teaspoons caraway seeds
1½ cups chopped bell peppers, preferably red	1 tablespoon dill seeds
2 cups chopped tomatoes	1 teaspoon black peppercorns
3 tablespoons red wine vinegar	4 cloves garlic, halved
3 quarts water, or enough to cover	2 teaspoons fresh thyme leaves, or 1 teaspoon dried thyme
2 cups sliced mushrooms	2 tablespoons chopped fresh basil, or 1 tablespoon dried basil
1 cup coarsely chopped cabbage	3 tablespoons soy sauce, optional
1¼ cups sliced parsnips	

In a stockpot, over medium heat, brown onions in oil about 15 to 20 minutes. The browned onions give a rich color and a slight sweetness to the stock. Stir in carrots, celery and peppers. Cover and cook over lowest heat for 10 minutes.

Add tomatoes and vinegar. Scrape the bottom of the pot thoroughly to loosen browned bits. Add water, mushrooms, cabbage, parsnips, potatoes and leeks. Cover and simmer until vegetables are soft, about 40 minutes. Add bay leaves, caraway and dill seeds, peppercorns and garlic. Cook for 1 hour more.

Cool slightly and strain through a colander, pressing lightly to extract all the juice.

Return broth to heat, uncovered, and cook at a slow boil until volume is reduced to 1½ quarts. Remove from heat and stir in thyme, basil and soy sauce.

Cool and freeze in ice cube trays. When frozen solid, transfer the cubes to freezer bags. One cube will flavor 1 cup of liquid.

YIELD: 42 cubes

30-MINUTE STOCK

A delicious, full-flavored stock that's quick to prepare. It can be frozen for 12 months.

1 small onion, coarsely chopped	⅛ teaspoon ground black pepper
1 clove garlic, minced	4 cups water
2 tablespoons vegetable oil or butter	1 pound chicken legs, or ¾ pound flank steak
1 teaspoon whole wheat flour	

In a large saucepan, sauté onions and garlic in oil for 2 minutes. Stir in flour and pepper, and sauté until lightly browned. Stir in water slowly to avoid lumps.

If you are making the stock with chicken, remove the skin from the chicken legs and discard. Cut the meat off the bones and cut it into bite-size pieces. Add meat and bones to the saucepan and simmer for 25 minutes. Cool stock and remove the bones. Skim excess fat from the surface or refrigerate overnight and lift off congealed fat. You can use the stock with the chicken meat, or remove meat before freezing to save for another purpose.

If you are making the stock with flank steak, cut the meat into bite-size pieces (if using frozen flank steak, thaw the steak only enough to ease cutting). Sauté the meat in a skillet over medium-high heat until browned. Add to the saucepan, cover and simmer for 25 minutes. Cool the stock. Skim excess fat from the surface or refrigerate overnight and lift off congealed fat. You can use the stock with the steak, or remove meat before freezing and save it for another purpose.

YIELD: 1 quart

GREEN VEGETABLE STOCK

Can be frozen for 12 months.

4 cups coarsely chopped
   onions
2 cups coarsely chopped
   carrots
2 cups coarsely chopped
   celery
1 cup coarsely chopped
   parsnips
   tops from 2 bunches of
   scallions, coarsely
   chopped
12-15 sprigs fresh parsley
4 large cloves garlic
1 bay leaf
3 whole cloves
12 black peppercorns
1 teaspoon fresh thyme
   leaves
4 quarts water

Place all vegetables, greens and seasonings in a stockpot. Add water. Bring to a boil, skimming off any foam that rises to the surface of the stock. Reduce the heat and let the stock simmer for 1 hour.

Strain through wet cheesecloth, discarding the solids. Cool and freeze.

YIELD: about 3 quarts

BROWN VEGETABLE STOCK

Can be frozen for 12 months.

4 cups coarsely chopped
   onions
2 cups coarsely chopped
   carrots
1 cup coarsely chopped
   parsnips
2 pounds coarsely chopped
   mushrooms
2 bunches scallions,
   coarsely chopped
12 sprigs fresh parsley
1 cup coarsely chopped
   tomatoes
4 cloves garlic
1 bay leaf
3 whole cloves
12 black peppercorns
1 teaspoon fresh thyme
   leaves
5 quarts water

In a large roasting pan, broil onions, carrots, parsnips and mushrooms, stirring so that all sides are browned.

Add the browned vegetables, along with scallions, parsley, tomatoes and seasonings, to a stockpot. Use 2 cups of the water to deglaze the roasting pan, scraping the pan well to loosen any browned bits. Add to the stockpot along with the remaining water. Bring to a boil, skim foam off the top, and then reduce heat to simmer, with the lid tipped slightly. Skim the foam.

After 1 hour, strain the stock through wet cheesecloth; discard the solids. Cool and freeze.

YIELD: about 4 quarts

BROILED TOMATO SOUP

This is a wonderful soup made with fresh or frozen herbs. Substituting dried herbs will greatly reduce its good flavor, so we suggest you do not use them. This soup can be frozen, without the cream, for 12 months.

2 cups coarsely chopped fresh tomatoes	ground white pepper, to taste
¾ cup chopped onions	½ cup tomato paste, thawed if frozen
1 cup chopped, mixed fresh or frozen herbs (oregano, dillweed, basil, parsley), thawed if frozen	2 cups Chicken Stock (page 73), thawed if frozen
½ cup butter	1 cup heavy cream
	½ cup grated Parmesan cheese

Cook tomatoes, onions, herbs and butter in a medium-size saucepan over medium heat until onions are soft. Stir often.

Add pepper and tomato paste and heat the mixture thoroughly. Pour the mixture into a blender or food processor and puree. You can freeze the soup at this point and add the stock, cream and cheese right before you serve it.

If serving the soup immediately, return the puree to the saucepan. Add the stock and heat through. Beat the cream until thick and then fold in the Parmesan cheese.

To serve, ladle soup into heatproof bowls. Top each bowl with a spoonful of the cream and cheese mixture. Broil until golden.

YIELD: 4 to 6 servings, 5 cups

WINIFRED GLASS'S TOMATO SOUP

This creamy soup was a winner in *Organic Gardening's* 1983 Gardener's Kitchen Recipe Contest. Don't freeze this soup.

1 quart peeled, quartered frozen tomatoes	3 tablespoons butter
½ teaspoon freshly ground black pepper	3 tablespoons whole wheat pastry flour
1 teaspoon minced fresh basil, or ½ teaspoon dried basil	1½ cups milk, or more to suit your taste
1 medium-size onion, finely chopped	2 teaspoons soy sauce

Put tomatoes in a medium-size saucepan. Add pepper and basil, cover and bring to a simmer. Chop up tomatoes slightly as they thaw. Cook until tomatoes are tender, about 15 minutes.

Meanwhile, in another pan, sauté onions in butter until slightly browned. Remove from heat and mix in flour and then milk. Stir a ladleful of the hot tomato mixture into the milk, then gradually stir this into the tomatoes. Add soy sauce, heat only until hot (do not boil) and serve.

YIELD: 4 to 5 servings, about 5 cups

NOTE: If you want to use tomatoes that have been frozen whole, peel them, thaw slightly, cut in quarters and measure out 3 cups. Then proceed as directed in the recipe. The smaller measurement compensates for the volume lost after thawing. Whole frozen tomatoes can be peeled easily by running them under cold water to loosen the skins.

CHAPTER 5

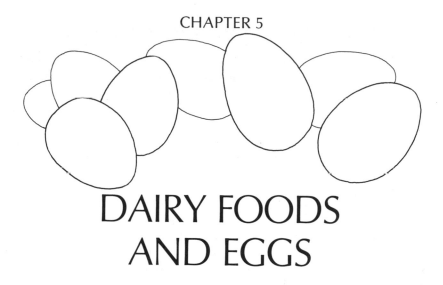

DAIRY FOODS
AND EGGS

Eggs and dairy products are not seasonal; therefore, storing them in the freezer is not necessary in order to have them year-round. Because dairy products are the very foods that perish most quickly, and certainly need refrigeration, you would think that they would be even happier in the freezer. But by and large, they are not. Not even ice cream, the ultimate freezer food, can stay in the freezer for long without an unfortunate change in its texture taking place. Nice as it would be to keep milk, cream, cheese and eggs on hold in the freezer, you can do so with only limited success. Life can be perverse, and so can food. Nevertheless, there are many methods you can employ to trick dairy foods into resting content in the freezer until you have need of them.

Cheese and Yogurt

Fresh cheese, unlike aged cheese, must be eaten soon after it is produced because it ripens and deteriorates rapidly. To retard its ripening, fresh cheese should be kept in the refrigerator, but not in the freezer, and consumed quickly. The category of fresh cheese includes all the mild-flavored, unripened, uncured cheeses. The softer a cheese is, the higher is its moisture content and the more drastically its texture deteriorates in the freezer.

87

Cream-style cottage cheese, which is up to 80 percent water, crumbles when it is frozen. However, both creamed and uncreamed cottage cheese, as well as ricotta cheese, will last up to a month in the freezer in their plastic containers. You can freeze your own homemade cottage cheese if you do not rinse the remaining whey when you remove it from the cheesecloth. In its unwashed form, cottage cheese will freeze satisfactorily.

Even though cream cheese loses its smoothness, you can still freeze it for up to four months and use it for cooking. Sour cream and yogurt can be frozen for a month, although they will separate in the freezer and will be suitable only for cooking when you thaw them. They will thaw in the refrigerator at a rate of eight hours per half pint. Farmer cheese, pot cheese, Neufchatel, petit Suisse and Gervais will not freeze well either. It's better to eat them fresh.

You will have the most success freezing fresh cheese and yogurt if you do so by combining them with other ingredients. Then they freeze admirably. For instance, you can mix cream and cream cheese in a dip and freeze it, and all will be well. Freeze yogurt mixed with a small amount of fruit preserves, and it will last in the freezer for two months. The smaller the ratio of cheese or yogurt to other ingredients, the better it freezes. If you want to experiment with freezing a fresh cheese on its own to see if you can get tolerable results, try a small portion, and leave some headspace at the top of the container. By the way, fresh cheese in the form of cheesecake freezes just fine.

Aged Cheese

There are about five dozen cheeses, ranging from soft to hard, that are commonly available commercially, and they react about five dozen different ways in the freezer. Generally speaking, any aged cheese tends to get crumbly and gritty when it has been frozen. If you only want to cook with the cheese you freeze, the crumblies won't be a problem. Cheeses that melt nicely before they're frozen will still melt nicely after freezing.

Though no cheese will escape some change in texture in the freezer, some will emerge in better shape than others. These are the semifirm to firm cheeses,

---

In 1911, Captain Scott left behind some Edam cheese when he departed from Antarctica, where winter temperatures hover around −60°F. In 1955, visitors to the South Pole found Captain Scott's camp still in sound shape. They reported that his Edam cheese tasted delicious.

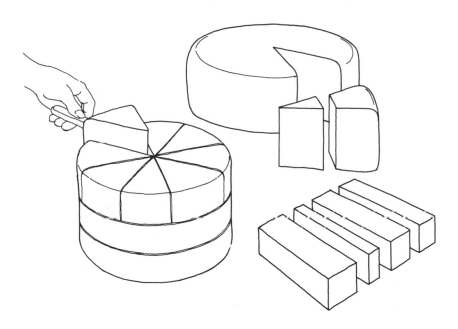

Large cuts of cheese can be divided for easier freezing. For large cylinders, slice crosswise with a cheese wire or thin knife into three equal parts, then cut each part into small wedges. Blocks or rectangles can be divided into smaller pieces, sized for future use, and thick wheels can be cut into wedges.

particularly brick, cheddar, Muenster, Camembert, Port du Salut, Swiss, longhorn, provolone, mozzarella, Liederkrantz, Parmesan and Romano. The latter two, in fact, should always be kept in the freezer rather than the fridge, if they're grated, in order to better retain their flavor and retard the formation of mold. Firm and semifirm cheeses can be frozen for two months.

Sometimes Limburger, Gouda, club and colby cheeses freeze well, but sometimes they don't. It depends on the kind you buy. No two natural cheeses are exactly the same. No two people find the same results equally acceptable. For that matter, no two freezers function with the same degree of efficiency, and that can influence how the cheese freezes, too.

Blue-veined cheeses are extremely sensitive to temperature variations, and you should freeze them only if you actually want them to come out crumbly so you can sprinkle them over salads. If that's your intention, go ahead and freeze Roquefort, Gorgonzola, American blue, Danish blue and any of the other blues, for up to three months. But if you want the cheese to retain its texture, to stay in one piece, don't freeze it. By the way, it's unfair to a good blue-mold cheese to actually mix it into a dressing that will all but obliterate its distinctive flavor. Rather than stir blue cheese into a mayonnaise, vinaigrette or sour cream dressing, crumble it over the salad greens before you add the dressing. Good news for Stilton lovers: It may be an exception to the blue-cheese rule. If you are lucky enough to have a wheel of Stilton, and you want to slice it and freeze it to serve after dinner occasionally, you may find that it actually thaws creamy rather than crumbly, even after several months, possibly because it is 55 percent butterfat.

Shredded and grated cheeses, as you might have guessed by now, freeze best (much better than thin slices or small cubes). They go into the freezer shredded or grated, they come out shredded or grated. No surprises. Shredded cheese will not pack together when it freezes; it'll stay loose and be just right for adding to a sauce or using as a topping for a casserole straight from the freezer. There is one caution, however: Don't add shredded cheese to the top of a casserole you are putting in the freezer. Handled in that way, the cheese will get soggy. Instead, when you reheat the casserole, top it with either fresh or frozen cheese at the last minute, right before it goes in the oven.

With all these caveats, you might wonder why anyone would want to bother to freeze cheese at all. Generally, the cheeses that do best in the freezer—the drier and more aged ones—also last longest in the refrigerator, which suggests that there is little need to freeze them at all. But if you belong to a food co-op and buy cheese in large quantities at low prices, or if you go shopping occasionally in faraway cities for unusual or favorite kinds of cheese, you may well want to stash the booty in your freezer.

How to Freeze and Thaw Cheese

First, to get the best results, freeze cheese in small portions, no more than ½ pound at a time. Pieces should be no more than ½ to 1 inch thick. Portions any larger or thicker take longer to freeze and therefore allow more time for ice crystals to form and break down the structure of the cheese. You should freeze cheese quickly so it won't get so crumbly. That means that you can't freeze a whole wheel of cheese.

It is crucial to wrap cheese very tightly for the freezer. If air reaches it, it dries out and is ruined. You can freeze an unopened package of store-bought cheese for up to two months. If you've opened the package and used some of the cheese, wrap the rest very tightly in aluminum foil and freeze it for no more than six weeks. Package grated cheese or cheese scraps in plastic freezer bags and, of course, get as much of the air out as possible.

If the cheese is a particularly pungent one, like the notorious Limburger, make sure that its odor doesn't escape the wrapping. Keep covering it until you can't smell it anymore, or the whole freezer will soon be redolent of Limburger cheese.

Frozen cheese may look blotchy and have odd streaks and colors, but don't worry—it's not your fault. The streaks will go away when the cheese thaws. Always thaw cheese in the refrigerator, because the more slowly it thaws, the less gritty it becomes. Cheese thaws in approximately six to eight hours. Keep it in its freezer wrappings as it thaws to help it retain its flavor and moisture. Don't try to refreeze cheese—it won't work. Instead, use it quickly. If you are cooking the cheese, handle it carefully. Cook it only briefly, and gently. High heat toughens the proteins and draws out the fat content, turning the cheese rubbery and making the dish it adorns oily.

Process Cheese

Process cheese is the bland cheese food you see in every supermarket. It is not aged, but rather manufactured to achieve uniform taste and a long shelf life. It is usually made with cheddar cheese, which is pasteurized and to which spices, emulsifiers, salt, coloring, cream and any of a number of other flavorings may be added. The pasteurization stops the bacterial growth that would cause the cheese to age. The other ingredients are intended to imitate the texture and flavor that real cheese develops naturally from aging and beneficial bacteria. The preservatives it contains give it a long shelf life. Process cheese is designed to last a long time in the refrigerator, so it doesn't really need freezing. But if you do choose to buy it and hold onto it, it will last five months in the freezer.

Butter

Butter is not temperamental in the freezer—far from it, as long as it's pasteurized. Margarine and butter are actually better off kept in the freezer if you're not going to use them right away. Margarine will freeze for five months. Unsalted butter can stay in the freezer for nine to twelve months, salted butter for six months because the salt makes the fats turn rancid more quickly. Clarified butter freezes well, and so does homemade butter made with home-produced pasteurized cream. Butter made with unpasteurized cream, however, will turn rancid almost immediately in the freezer. If you make sure to thoroughly rinse your home-churned, unsalted butter of its buttermilk, it will keep in the freezer for nine months.

Butter doesn't just flavor vegetables and fruits. Fruits and vegetables can flavor butter, too. And so can herbs and spices and syrups, and many other foods as well. The more flavors and colors of blended butter you stock in the freezer, the wider the range of interesting possibilities you can draw on when you cook.

You can freeze homemade butter in freezer containers in ½- to 1-cup quantities, or mold it and wrap it. All butter needs to be carefully wrapped because it easily picks up other odors in the freezer. If you're freezing store-bought butter, you can keep it in the original carton, but overwrap it with freezer paper. Naturally, you should thaw all butter in the refrigerator (it will take about four hours per half pound).

The plethora of butter concoctions in this chapter suggest the versatility butter lends itself to. Flavored butters can not only be made with fresh herbs or shellfish, the classic combinations, but they can also be concocted from almost any other food that you can mash or grind up.

Try making vegetable butters, for instance, not only as a tasty way to use fresh vegetables but also for the interesting color they add to a meal. Asparagus butter, for example, is a lovely light green color that blends well in color and flavor as it melts over cooked asparagus. You can make it by blending softened butter together with the woody stems of asparagus that have been cooked until tender with a few garlic cloves. You can cook carrot slices in the same way, and add them to the blender with softened butter and a few herbs. The resulting orange butter adds character and color to potatoes. Try roasting or sautéing sweet red peppers, pureeing them, then mixing the puree with butter.

Summer vegetables and herbs can go into compound butters that will accompany meals all winter. Try a sorrel and parsley butter for fish, a sage and shallot butter for poultry or game, a basil butter for tomatoes. It's a matter of finely chopping the herbs in question and blending them with softened butter in roughly equal amounts. The addition of a few drops of lemon juice seems to enhance their keeping qualities in the freezer.

It's amazing what you can do with butter. Try creaming it together with maple syrup or honey, and keep it in the freezer to use on pancakes or waffles. Make butter curls, swirls, balls and shapes of all sorts, and keep them frozen in a freezer bag or container. All these small fantasies enhance special-occasion meals out of all proportion to their size or the time it takes to create them. If your life is full of unexpected visitors, compound butters in the freezer can instantly elevate the quality of a hastily prepared meal.

Milk

You can freeze milk, but you may not be happy with the results. It will thaw with little flecks. Some people don't mind the flecks or the taste. They drink milk from the freezer and think it's all right; others don't care for the taste at all. Goodness knows, food tastes are a personal thing.

If you keep a milk cow or a goat, you can freeze the excess milk—whole, skimmed, pasteurized or raw. Milk of any description, including buttermilk, will last in the freezer for a couple of months, but quality will begin to fade after one month. You can freeze milk in the carton you buy it in, or in a straight-sided plastic container. Just make sure you leave 1 inch of headspace for pints to expand, 1½ inches for quarts and 2 inches for half gallons. The quart size is best for freezing purposes, because anything larger thaws so slowly you may give up on it. Make sure, too, that the container is closed very tightly, because freezer odors will readily permeate milk and ruin the flavor. Most dairy products are particularly susceptible to absorbing odors and flavors from the foods they are in contact with. That's desirable if you're making a glass of chocolate milk, but if you're keeping milk products in the freezer, it's a problem you must guard against with careful wrapping.

Thaw milk only in the refrigerator. It takes eight to ten hours per pint. At room temperature, it will thaw in about six hours, but don't sacrifice safety for speed. Milk left at room temperature will undergo lactic acid fermentation, and many kinds of bacteria can grow in it.

Cream

Cream that contains less than 40 percent butterfat, and that includes light cream, doesn't freeze well because the butterfat separates out in the freezer. If you do freeze it, light cream won't look appetizing in coffee or on cereal, because the flecks of fat that have separated out float on the surface, but you can cook with it. Just beat it to incorporate the butterfat from the top. Then add it to soup, gravy or custard. It will keep in the freezer for about two months, and will thaw in the refrigerator at a rate of eight hours per half pint (half-and-half can be frozen and thawed for the same length of time).

If you have goat's cream or cow's cream from your own goat or cow, you can freeze it if you want to hold it until you can build up a big supply for butter making. Pour each day's complement into its own plastic container, which has been sterilized with a commercial dairy sterilizing solution. Leave 1 inch of headspace for expansion, and fit the lid on tightly to keep out freezer odors.

Whipped Cream

If you freeze heavy cream before you whip it, it will be grainy and separated when it thaws. Whipping it, if it whips at all, will take longer and produce highly variable results, although adding a few drops of lemon juice should help. However, you can freeze heavy cream for two months. It will thaw in the refrigerator at a rate of eight hours per half pint.

It's a better idea to whip cream before you freeze it. To get the most volume, start off by putting the bowl, beaters and cream into the freezer for ten minutes. Whip the cream, flavor it and then freeze it in small portions.

If you've whipped the cream until it forms peaks, you can dollop it onto a cookie sheet lined with freezer paper. Put the cookie sheet in the freezer. When the mounds of cream can be moved without damaging them, store them in a freezer bag or rigid container. If you use a cake decorator or cookie press, you can mold the cream into some neat swirls and shapes. Use your frozen cream whimsies to enrich and decorate the tops of desserts, cakes, pies or hot chocolate (where it certainly beats marshmallows). The little dabs of cream will thaw in a scant ten minutes, so pop them straight from the freezer to their intended destination. In the freezer, they'll stay in top shape for three months.

Pipe little mounds of whipped cream onto a cookie sheet and place the sheet in the freezer. When the mounds are firm, remove from the cookie sheet and place in a plastic freezer bag. To use, remove the number of mounds you need and place them directly on the cake, pie or dessert.

Eggs

Though an egg looks simple enough, it isn't, and its complexity makes it tricky to freeze. The yolks and whites are distinctly different from each other in composition, for one thing, and they react differently to freezing. The yolk is the repository for all the fat in the egg, almost all its vitamins, all the cholesterol and most of the calories. The white is mostly water and pure protein. Raw egg yolks don't freeze particularly well, but raw whites freeze fine. Hard-cooked yolks freeze perfectly, but hard-cooked whites freeze poorly.

How well cooked dishes freeze when they contain eggs appears to depend on the function that eggs take on in the recipe. If the eggs are in the recipe to moisturize, aerate or flavor the food, then they will freeze well. That is why French toast, potato pancakes, quiche and cakes freeze well when they are made with eggs. (Quiche, for example, will freeze for three months, and thaws in the refrigerator in 12 to 24 hours.) But when the eggs are intended to act as an emulsifier, that is, to keep together a mixture that might otherwise separate, then the dish will not freeze at all well. Mayonnaise, which is predominantly egg yolks and oil, doesn't freeze successfully, and neither do egg custards or egg-based sauces. Egg yolks simply break down as emulsifiers at extreme temperatures, either high heat or freezing cold, and cease to work.

Though freezing does not significantly affect the texture of raw egg whites, it makes cooked egg whites rubbery. Keep stuffed eggs, egg salad and egg white garnishes from ever seeing the inside of your freezer.

Freezing Whole Eggs

Don't freeze whole raw eggs in the shell, because the shell will crack when freezing causes the contents to expand. You can, however, crack the eggs open and freeze the yolks and white mixed together, or freeze the yolks and whites separately.

If you see eggs on sale for 45 cents a dozen, you can buy an extra carton to freeze for six to nine months. (Large eggs are usually the best size to buy because they contain the greatest volume of liquid for the price.) Crack the eggs into freezer containers in recipe-size portions and stir them together gently. Don't beat them, or you will produce air bubbles that will dry out the eggs in the freezer.

Salmonella bacteria find their way into eggs when the shells are cracked open. The contents are protected only as long as the shell remains intact. To be safe, don't freeze eggs that have cracks, and get your containers of eggs into the freezer quickly.

It's important to thaw eggs only in the refrigerator, even though a pint will take 10 hours to thaw. Countertop thawing will increase the salmonella population if the eggs reach room temperature. You can, however, keep the freezer container in cold water for 3 hours, to speed the thawing. Use the eggs within 24 hours after thawing.

It takes about five large eggs to make up 1 cup. One whole egg equals about 3 tablespoons of stirred yolk and white.

Because of the strange workings of salmonella bacteria, which love uncooked eggs, there are certain precautions you should take when you use eggs that you

---

Eggs are considerably more complex than they look. Predicting their behavior in the freezer depends on knowing what their function is in the food you are freezing.

---

have frozen. It's advisable to use the eggs in making long-cooking recipes like cakes and breads, rather than in fast preparations like scrambled eggs, omelets or hollandaise sauce, or raw in blender milk shakes. Salmonellae find their way into an egg when you crack it open without washing it first. If you freeze and thaw the egg, they will multiply very rapidly if the egg reaches room temperature. A quick stint in the pan when you make a fried egg will not provide sufficient heat to destroy the bacteria.

Only if the eggs you freeze are new-laid, scrubbed clean with soap and water before cracking, and frozen immediately can you use them raw or in a quick-cooking preparation.

Freezing Egg Whites

If you make ice cream, you probably have a lot of leftover egg whites. You can freeze them to use later. Since they're mostly protein, egg whites freeze with ease. Drop each white into a cup of an ice cube tray. Or freeze the whites together in a quantity that will accommodate a recipe you like. If you freeze more whites together than you can sensibly use, you will have a rough time hacking off the amount you need. Add whites to a freezer container, and stir them together gently to avoid having air bubbles form, which will dry them out. One egg white equals 2 tablespoons of stirred egg whites. Eight egg whites equal 1 cup of stirred whites. The whites will store for nine to twelve months, and thaw in the refrigerator at a rate of nine hours per cup. Use them to make sherbet, baked Alaska and meringue desserts.

Freezing Egg Yolks

If you make angel cake or meringues, you'll have egg yolks aplenty to contend with. You can't just stir the yolks together and freeze them the way you can egg whites. Freezing makes egg yolks pasty and unmanageable unless you stabilize them first. Honey is an effective way to keep yolks from turning to a gel. Add 1 tablespoon of honey for every three eggs you freeze, and note on the label that the yolks contain honey.

If you want to freeze the yolks in ice cube trays after you've stirred them together, 1 egg yolk is the equivalent of 1 tablespoon of stirred egg yolk. Twelve stirred yolks equal 1 cup. Egg yolks will store for six to nine months in the freezer—not as long as whites because the yolks are higher in fat content. A cup of egg yolks will thaw in the refrigerator in about nine hours.

Ice Cream

Ice cream represents the finest hour of dairy foods in the freezer. It starts with custard that takes a half-hour of slow cooking and stirring to produce. The addition of cream and flavorings follows . . . and that's basic ice cream, though the variations are legion. The recipe for All-Cream Ice Cream (page 108) in this chapter is a variation that omits the custard and is made with just cream.

The churning of the ingredients to freeze them usually takes place a day after the mixture is made. The wait serves to increase the yield and smooth out the texture of the ice cream, as well as allowing ample time for the mixture to cool and harden. But note that ice cream made with honey instead of sugar will be softer in consistency.

Even before there were freezers, there was ice cream. The first hand-cranked ice cream machine was invented by Nancy Johnson in 1846. Her invention is still the basis of the hand-cranked and electric ice cream makers we use. The operating principle is based on the fact that salt lowers the temperature at which water turns to ice.

A wood bucket containing a metal pail surrounded by a layer of ice and rock salt is the typical configuration. The more salt you add to the ice, the lower the temperature drops, from 32°F, the freezing point, down to as low as −5°F. As the salt-ice mixture brings down the temperature to create ice cream inside the metal pail, you turn a paddle, or dasher, by hand or machine crank. The cranking prevents the ice cream from crystallizing as it freezes.

If you did not churn the mixture, but simply froze it, it would turn into a solid block of ice. The churning keeps interrupting the ice masses that are trying to form. The more you churn, the longer the freezing takes and the tinier are the ice crystals that can form. Really smooth ice cream, in fact, has very tiny ice crystals.

Another function of the stirring is to let air into spaces where ice crystals would like to be. The incorporation of air expands the volume of the ice cream by about 25 percent, and makes it light. Commercial ice cream has a lot more air pumped into it than that, although federal standards keep down the quantity of air that can be whipped into it. A gallon of commercial ice cream must weigh no less than 4½ pounds. By comparison, a gallon of homemade ice cream normally weighs between 6 and 8 pounds. That's why homemade ice cream is denser and less smooth than the ice cream you buy.

Hand cranking ice cream is sometimes faster than machine cranking, but of course it's much more arduous, particularly during the last five minutes. While hand cranking a batch of ice cream can provide entertainment for the whole family, ice cream making is a one-man job with an electric ice cream maker. Add the cool custard to the freezer can, layer ice and rock salt around it, and turn on the machine. Add more salt and ice as it melts, and you have ice cream in around 30 minutes—about the same length of time that hand cranking usually takes.

There are several kinds of ice cream machines, including some models that fit right in the freezer, which does the freezing instead of salt and ice. Meanwhile, the machine does the churning. Experiments with hand-cranked and electric ice cream makers at the Rodale Test Kitchen determined that all these machines make good ice cream. Hand-cranked ice cream is marginally smoother and creamier, and it had its advocates during our tasting sessions. But so did the slightly grittier electric ice cream maker version.

Still-Freeze Ice Cream

This is *the* freezer ice cream. It needs no salt or ice, because the freezer chills the mixture, which is usually packed into an ice cube tray without dividers. Sometimes the ice cream comes out of the freezer several times to have more air whipped into it. It requires only ordinary kitchen tools to make, and it's very smooth and delicious.

Still-freeze ice cream is distinctly different from churned ice cream in the preparation and in the final product. Instead of churning it to prevent it from crystallizing, you use an emulsifying agent instead, usually eggs, cornstarch or gelatin.

There are recipes in this chapter for still-freeze ice cream, and several for still-freeze ices. The latter do not require repeated removal from the freezer to beat more air into them. Instead, you need only puree fruit in the blender or food processor along with an egg white or some milk or gelatin, until it reaches a sherbetlike consistency. ·

Don't make these quick ices to store in the freezer. Instead, plunder the freezer to find the fruits to make them, then eat them quickly. Ices can be made from just about any fruit or vegetable that is combined with a dairy product or gelatin in an icy, frothy and quick-to-prepare way.

Ices are so fresh-tasting and low in calories that they have become a cornerstone of the new light cuisine. They are, however, very old. Among the other treasures Marco Polo brought back from China was a recipe for ice milk. That was in the thirteenth century. He knew what was good.

Storing Ice Cream in the Freezer

Store still-freeze ice cream for two days in ice cube trays that have dividers you can remove. The large amount of surface area speeds up the freezing process considerably. Cover each tray tightly with the tray's lid or freezer plastic or foil.

Store churned ice cream for up to six weeks in a tightly wrapped container. You can press aluminum foil or plastic wrap over the ice cream to block off the air between the ice cream and the lid. Ice cream can readily pick up the flavors that linger in your freezer containers. Either keep it in new containers earmarked for ice cream, or really scrub the containers you do put it in.

Commercial ice cream will last in the freezer in top condition for no more than two months. You can keep it a little longer if you repack it in freezer containers, or overwrap it in a freezer bag. Even during that time, the ice cream will shrink in volume and lose its nice texture if it melts even slightly and then refreezes. It's important, therefore, to keep it in your freestanding freezer rather than in the freezer compartment of your refrigerator.

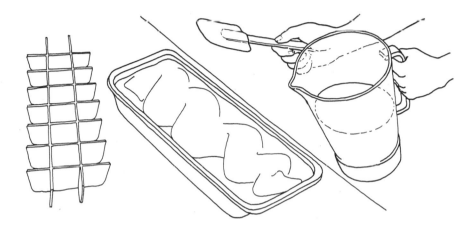

Use ice cube trays with removable dividers to still-freeze your ice cream.

Serving Ice Cream

Homemade ice cream needs time in the freezer to firm up. When you serve it, however, it should be "ripe," which means it is firm but not rock hard and at a temperature between 8° and 12°F. Bear in mind that ice cream made with honey will be softer than ice cream made with sugar. Before serving, ripen the ice cream by removing it from the freezer and holding it in the refrigerator where it can soften uniformly without simply getting soft around the edges. How long the ripening takes depends on the container your ice cream is in, and how cold your freezer is.

To test if your ice cream is ripe, plunge a knife into it. It should have some difficulty passing down to the bottom but it should be able to get there. If it doesn't reach the bottom because the ice cream is too hard, return the ice cream to the refrigerator for a few minutes. Ice cream pies and cakes, on the other hand, should ripen at room temperature.

BETTER BUTTER

Although calories will be about the same, the saturated fat level is substantially reduced by blending butter and vegetable oil together. The flavor is still quite satisfying, and Better Butter is superb for cooking—butter flavor with less tendency to burn or stick than straight butter. Better Butter can also be used as a base for herb butters.

We recommend doing 2 pounds at a time and freezing the excess, since cleanup time is involved. Better Butter can be frozen for 4 months.

2 pounds butter, at room
    temperature

2 cups sunflower or
    safflower oil

Blend together butter and oil thoroughly in a food processor or blender. Store in small plastic or glass containers. Soft margarine containers are a good choice. Freeze any butter that will not be used within 2 weeks.

YIELD: 6 cups

BASIC HERB BUTTER

The proportion of butter to herbs can be varied to suit your purpose. To simply make a flavored butter, use ½ cup butter to ¼ cup finely chopped herbs. If, however, the objective is to preserve the herb flavor without adding other flavor to the herbs, use just enough butter to coat the herbs and make them stick together, about ½ cup butter to ¾ cup finely chopped herbs. This more intense herb butter can later be cut with additional fresh butter. Always use top-quality, freshly purchased butter when making flavored butters for freezing.

Herb butters can be frozen for 9 months.

½ cup butter, at room
    temperature
¼–¾ cup finely chopped fresh
    herbs, stems removed

2 teaspoons lemon juice,
    optional

Place butter in a bowl and cream it completely with a fork or electric mixer. Add herb(s) and stir until completely mixed. Stir in lemon juice, if desired. Freeze in ice cube trays, by shaping into logs, or in small containers such as 1-cup jelly jars.

YIELD: ¾ to 1¼ cups

NOTE: Herbs may be chopped in a food processor, using on/off turns, and the butter added to the bowl when the herbs are fine enough. The cut will not be as even as hand-chopped herbs, and the butter will be greener, as more juices are freed. But the food processor method is quick and convenient for making large batches.

MUSTARD GREEN GARLIC BUTTER

The tangy leaves of the Oriental mustard greens are delicious in butter. Serve a pat with fish, cooked vegetables, broiled meats or to make an extra-special garlic bread. The butter can be frozen up to 4 months.

½ cup very finely chopped mustard greens, stems removed
½ cup butter, at room temperature

1 large clove garlic, minced
1 teaspoon lemon juice

In a small bowl, combine mustard greens, butter, garlic and lemon juice until well blended. Chill until ready to use, or freeze.

YIELD: ¾ cup

NOTE: Adding a vegetable to garlic bread gives it a bit of nutritionally redeeming value and an intriguing flavor. To make mustard green garlic bread, slice a small loaf of French or Italian bread into thick slices, not cutting quite all the way through. Spread each slice with a thick layer of Mustard Green Garlic Butter. Wrap the loaf with foil, leaving a vent in the top. Bake for 15 minutes at 375°F until steamy hot.

VARIATION: For a simpler approach, stir 3 tablespoons prepared mustard into ½ cup butter.

CHEESE BUTTERS

In addition to herbs, many other flavors can be added to butter and kept frozen, to use with vegetables, as fancy sandwich spreads or melted onto hot biscuits and muffins. Freezing time will be related to the added ingredients but averages 6 months.

Cheese Butter with Semi-Hard "Table" Cheese

1 cup butter, at room
    temperature
½ cup finely shredded
    cheese, either sharp
    cheddar, English
    Cheshire, Jarlsberg,
    or fontina

2 teaspoons caraway,
    fennel or dill seeds,
    optional
1 teaspoon paprika,
    optional

In a medium-size bowl, mash butter with a fork or electric mixer. Stir in chesse, seeds and paprika until thoroughly mixed. Pack tightly into glass jars. Will hold in refrigerator for 2 weeks. Freeze any extra butter.

YIELD: 1¼ to 1½ cups

Cheese Butter with Strongly Flavored Cheese

1 cup butter, at room
    temperature
¼ cup crumbled blue
    cheese or freshly
    grated Parmesan or
    Romano cheese

In a medium-size bowl, mash butter with a fork or electric mixer. Stir in cheese until thoroughly mixed. Pack tightly into glass jars. Will hold in refrigerator for 2 weeks. Freeze any extra butter.

YIELD: 1¼ to 1½ cups

Cheese Butter with Soft Cheese

1 cup butter, at room
temperature
½ cup very soft Camembert
or Brie, with some rind
removed, or chèvre (a
soft creamy French
goat cheese)

2 tablespoons chopped
fresh herbs, such as
garlic chives, thyme,
basil or parsley,
optional

In a medium-size bowl, mash butter with a fork or electric mixer. Stir
in cheese and herbs, if desired, until thoroughly mixed. Pack tightly into
glass jars. Will hold in refrigerator for 2 weeks. Freeze any extra butter.

YIELD: 1¼ to 1½ cups

SPICE BUTTER

Good tossed with rice, pasta or beans. Spice butters can be frozen for
9 months.

1 cup butter, at room
temperature
1 tablespoon curry powder,
or 1 teaspoon ground
cinnamon, or ¼
teaspoon ground
cardamom

In a medium-size bowl, mash butter with a fork or electric mixer. Stir
in spice until thoroughly mixed. Pack tightly into glass jars. Will hold
in refrigerator for 2 weeks. Freeze any extra butter.

YIELD: 1 cup

VARIATION: Substitute a combination of ground nutmeg and cloves for
the curry or cinnamon.

DILL SCALLION BUTTER

Can be frozen for 9 months.

½ cup butter, at room
temperature
⅓ cup finely chopped fresh
dillweed, stems
removed
¼ cup finely chopped
scallions, including
some of the green tops

2 tablespoons chopped
fresh parsley
2 teaspoons lemon juice

Prepare as for Better Butter (page 100).

YIELD: ¾ cup

SWEETENED BUTTER

A very good spread on moist, dark bread. This butter can be frozen for 9 months.

1 cup butter, at room
temperature

3 tablespoons preserves or
jelly, or ¼ cup honey

In a medium-size bowl, mash butter with a fork or electric mixer. Stir in preserves, jelly or honey until thoroughly mixed. Pack tightly into glass jars. Will hold in refrigerator for 2 weeks. Freeze any extra butter.

YIELD: 1 to 1¼ cups

CITRUS BUTTER

Unsweetened, this butter is great with fish. Sweetened, it is good on toast and muffins. Citrus Butter can be frozen for 3 months.

1 cup butter, at room
   temperature
2 tablespoons grated or
   finely shredded orange
   or lemon peel

2-3 tablespoons orange or
     lemon juice
2 tablespoons honey,
   optional

In a medium-size bowl, mash butter with a fork or electric mixer, stir in orange or lemon peel, juice and honey until thoroughly mixed. Pack tightly into glass jars. Will hold for 2 weeks in refrigerator. Freeze any extra butter.

YIELD: 1¼ to 1½ cups

NUT AND SEED BUTTERS

This is a tasty and unusual addition to fish and vegetables. Freeze this butter up to 9 months.

1 cup butter, at room
   temperature

¼ cup finely chopped
   black walnuts

In a medium-size bowl, mash butter with a fork or electric mixer. Stir in nuts until thoroughly mixed. Pack tightly into glass jars. The butter will hold in refrigerator for 2 weeks. Freeze any extra butter.

YIELD: 1¼ to 1½ cups

VARIATIONS: Substitute ¼ cup hickory nuts or toasted sesame seeds (page 54), ½ cup finely chopped pecans, walnuts or hazelnuts or 2 tablespoons lightly ground caraway, fennel or dill seeds for the black walnuts.

FRUITY FRENCH FROZEN CUSTARD

This is a light, custard-based ice cream to prepare with an ice cream maker. You can use almost any fruit to flavor the custard—blackberries, blueberries, cherries, mangoes, melons, peaches, raspberries, strawberries. It can be frozen for 2 months.

1 pint light cream or half-and-half

2 eggs, separated

2 teaspoons vanilla extract, or 1 teaspoon almond, orange, lemon or mint extract (see Note)

½ cup honey

1 tablespoon gelatin, softened with ¼ cup water

13 ounces evaporated milk

½ pint heavy cream

2 cups lightly sweetened fruit, almost thawed if frozen, optional

Heat a double boiler while getting ingredients ready. Add light cream to double boiler and heat slightly. Lightly beat egg yolks, add to the light cream and beat until smooth. Stir in desired flavoring and honey. Add softened gelatin and blend well. Heat mixture to dissolve the gelatin, stirring frequently, but do not bring mixture to a boil. Remove from stove and let cool slightly. Beat the egg whites till soft peaks form, then fold them into the cream–egg yolk mixture. Chill mixture in refrigerator for several hours or overnight.

When thoroughly chilled, add evaporated milk to mixture and beat with an electric mixer until smooth. Add the heavy cream and beat again for about 1 minute.

Put the ingredients into the ice cream maker container and stir with a wooden spoon or rubber spatula. Be sure ingredients do not exceed the maxi-line level of the can, because the ingredients need room for expansion.

Meanwhile, finely chop fruit in a blender or food processor. When ice cream is finished (hard to churn), swirl fruit through ice cream with a spatula. Place in freezer at least 30 minutes before serving.

YIELD: ½ gallon

NOTE: Extracts may be used in combination, decreasing vanilla as others are used; different fruits go with different extracts. Strawberry is good with vanilla, cherry and peach with almond, melon with mint, and the tropical fruits with orange or lemon.

ICE CREAM PIE

Can be frozen for 2 months.

1¼ cups toasted shredded
    coconut
5 ounces cream cheese, at
    room temperature
3½ cups All-Cream Ice Cream,
    softened (page 108)

6 cups peach ice cream
2 cups pureed strawberries,
    thawed if frozen
1 cup pureed raspberries,
    thawed if frozen
6 whole almonds

Place a 10-inch deep-dish pie plate in refrigerator to chill.

In a medium-size bowl, stir together coconut, cream cheese, and 1 cup All-Cream Ice Cream until mixed thoroughly. Spread this mixture over the bottom and sides of the chilled pie plate. Place in freezer for about 15 minutes, or until firm.

Spread 2 cups peach ice cream over coconut layer. Place in freezer for about 15 minutes, or until firm.

Combine 2 cups All-Cream Ice Cream with the strawberry puree. Spread half of the mixture over peach layer and freeze until firm.

Spread 2 cups peach ice cream over strawberry layer. Freeze until firm. Repeat with remaining strawberry mixture and then the remaining peach ice cream, freezing until firm before adding the next layer.

Combine raspberry puree with remaining ½ cup All-Cream Ice Cream. Spread around outside edge of peach layer.

Place almonds in a star shape in the center of the pie. Cover with wax paper and foil and place in a large freezer bag. Freeze for at least 8 hours before serving. Allow to sit at room temperature 15 to 20 minutes before serving.

YIELD: 12 to 16 servings

VARIATION: Sprinkle ⅓ cup carob chips on top.

ALL-CREAM ICE CREAM

No cooking needed here—the maple syrup dissolves quickly in the cream, and ice cream is ready in about 45 minutes—start to finish. It can be frozen for 2 months.

2 cups heavy cream	½ cup maple syrup
2 cups half-and-half	2 teaspoons vanilla extract

In the ice cream maker container, combine the cream, half-and-half, maple syrup and vanilla. Assemble and operate ice cream maker according to manufacturer's directions.

When ice cream hardens, transfer to freezer container and freeze for at least 30 minutes before eating.

YIELD: about 5 cups

VARIATIONS: Add 1 cup chopped pecans, toasted almonds or toasted hazelnuts (page 54) to ice cream after it has thickened.

Omit 1 cup half-and-half and puree 2 cups peaches or strawberries in the remaining cup of the half-and-half. Add to ice cream maker container at the beginning of freezing.

For a fruit swirl, cook 2 cups chopped fruit (berries, cherries or apples) in ½ cup apple juice and ⅓ cup honey until soft. Mash and cool. Swirl through thickened ice cream.

ZUCCHINI ICE CREAM

When the weather is hot, your hunger for ice cream is great, and the zucchini vines are loaded, dare to try—ice cream from zucchini! Can be frozen for 1 day.

2 cups pureed raw or blanched (but not fully cooked) zucchini	2 eggs, beaten
	2 teaspoons vanilla extract
	pinch of ground nutmeg
½ cup honey	1 cup heavy cream

Heat zucchini and honey in a saucepan over medium-low heat.

Add 2 tablespoons of hot zucchini to eggs and mix well. Add egg mixture to hot zucchini mixture and cook until mixture thickens slightly.

Chill. Add vanilla, nutmeg and cream. Pour into ice cube trays and place in freezer until amost solid.

Place in chilled mixing bowl and beat until smooth with electric mixer. Freeze again until ready to serve.

YIELD: 1 quart

ICE CREAM CAKE

Soften the ice cream slightly by beating with mixer before beginning. The finished cake can be frozen for 2 months.

4 cups carob ice cream	1½ cups cookie crumbs
4 cups All-Cream Ice	4 cups strawberry ice
Cream (page 108)	cream

Chill a 9-inch springform pan in the freezer for 10 minutes.

Remove pan from freezer and spoon half of the carob ice cream into the bottom of the pan, spreading evenly. Cover with foil or plastic wrap and place in freezer for at least 45 minutes, or until firm.

Spread about 1 cup All-Cream Ice Cream over the top of the carob layer and sprinkle with ¼ cup cookie crumbs. Cover and return to the freezer for about 1 hour.

Next, place half of the strawberry ice cream over the All-Cream Ice Cream and spread evenly. Sprinkle with ¼ cup cookie crumbs. Place in freezer for 1 hour.

Place serving dish in refrigerator to chill.

Follow with another carob layer and strawberry layer following the same procedure, using ¼ cup cookie crumbs on top of each layer, and allowing each layer 45 minutes to 1 hour to set.

Remove cake from the springform pan onto chilled serving plate.

Spread remaining 2 cups All-Cream Ice Cream over the top and sides of the cake and top with remaining cookie crumbs. Cover and return to freezer until ready to serve. Allow to sit at room temperature 15 to 20 minutes before serving.

YIELD: 12 to 16 servings

NATURALLY DELICIOUS SHERBET

This recipe requires an ice cream maker. It can be frozen for 2 months.

2 cups unsweetened fruit of your choice
1 cup frozen apple juice concentrate

13 ounces evaporated milk, cold

Puree fruit in a food processor or blender. Add juice concentrate to pureed fruit and blend. Pour the pureed mixture into the container of the ice cream maker.

Pour evaporated milk into a large mixing bowl. With an electric mixer, beat milk until it is fluffy like whipped cream.

Pour the whipped milk into the ice cream container with pureed fruit. Stir all ingredients together. Process until thick. Place in freezer for 30 minutes before serving.

YIELD: ½ gallon

STRAWBERRY YOGURT ICE CREAM

Almost any kind of ice cream, even the commercial variety, is delicious, as well as more nutritious and less fattening, when a portion of yogurt is whipped into it. This ice cream can be frozen for 1 day.

1 cup heavy cream
3 cups strawberries, slightly thawed if frozen

1 cup yogurt
⅓ cup honey

Whip cream until stiff and set aside.

Place strawberries, yogurt and honey in a food processor or blender and process until smooth. Fold into the whipped cream, pour into ice cube trays and freeze.

When frozen solid, thaw slightly, then blend again in blender or food processor until smooth. Store in freezer containers.

YIELD: 5 cups

VARIATIONS: Substitute sliced peaches or pitted sweet cherries for the strawberries.

QUICK FROZEN YOGURT

Commercial frozen yogurt is often over-sweetened and full of various additives and stabilizers. Fortunately, hardly anything is simpler to make, tastier, or more nutritious than the homemade version. Can be frozen for 1 day.

2 cups blueberries or
   chopped strawberries,
   thawed if frozen

1/3 cup honey
2 cups yogurt

Cook berries and honey together for about 4 minutes. Cool completely.
   Place berry mixture and yogurt in the bowl of a food processor or blender and process until thoroughly blended. Freeze in ice cube trays until almost solid, then process again if smoother texture is desired. Refreeze.

YIELD: 3½ to 4 cups

NOTE: Frozen yogurt is good made in Popsicle molds.

FROZEN YOGURT POPS

These pops can be frozen for 1 day.

2 bananas, slightly thawed
   if frozen
2 cups yogurt
2 teaspoons vanilla extract

½ cup frozen pineapple
   juice concentrate,
   slightly thawed
1/3 cup honey, optional

Slice bananas into a blender. Add yogurt, vanilla and pineapple juice concentrate; blend until smooth. Taste and add honey if needed. Pour into Popsicle molds and freeze until firm.

YIELD: 6 to 8 pops

NOTE: Instead of making ice pops, you can simply pour the mixture into a freezer bowl, cover and freeze until firm. Before serving, let it "ripen" in the refrigerator for 30 minutes so it can be scooped out more easily.

VEGETABLE GARDEN OMELET

Use a colorful mixture of frozen vegetables for this omelet, such as cauliflower, carrots, peas and scallions.

1 tablespoon vegetable oil
1 cup coarsely chopped combination of frozen cauliflower, carrots, peas and scallions, partially thawed
¼ cup slivered almonds
2 tablespoons water
1 tablespoon butter
4 eggs, beaten

Heat oil in a large skillet and add vegetables and almonds. Stir a few minutes to coat with oil. Add water and cook until vegetables are hot.

Place vegetables in a bowl. Dry pan, add butter and return to heat. When butter is melted add eggs. After a few minutes add vegetables.

When eggs are almost set, fold omelet in half and then turn to make sure the inside is completely cooked. Slice and serve hot.

YIELD: 2 servings

PEPPER AND CHEESE CUSTARD

Can be frozen for 2 months.

1½ cups seeded, diced fresh or frozen sweet bell or frying peppers
⅓ cup finely chopped fresh or frozen hot chili peppers
1 teaspoon vegetable oil
2 cups large-curd cottage cheese
2 cups shredded cheddar, Swiss or Muenster cheese, about ½ pound
5 eggs
½ cup milk

Sauté sweet and chili peppers in oil until slightly tender, about 4 minutes if fresh, 8 minutes if frozen. Remove from heat.

Mix together cottage cheese, shredded cheese and sautéed peppers. Arrange cheese mixture in an oiled 8 × 8-inch baking pan.

In a bowl, beat eggs and milk together thoroughly. Pour the egg mixture over the cheese mixture. Bake at 350°F for about 35 to 40 minutes, or until golden on top, and a knife inserted in center comes out clean.

Thaw before reheating and bake in 350°F oven until hot, about 15 minutes.

YIELD: 4 servings

VARIATIONS: Replace part of the peppers with a cooked vegetable such as peas, drained chopped zucchini or corn.

OVEN-BAKED FRITTATA

4 scallions, including green tops, thinly sliced	2 tablespoons butter
2 cups frozen broccoli spears, green beans, sugar peas, sliced carrots or asparagus spears, partially thawed	¼ cup water
	½ cup milk
	1½ cups shredded Swiss or cheddar cheese
	6 eggs, beaten

In a 10- or 12-inch ovenproof skillet, sauté scallions and vegetables over medium heat in 1 tablespoon butter for a few minutes, then add water to pan. Cover and steam until vegetables are tender but still crisp, 4 to 10 minutes.

Uncover and cook away any remaining water. Add remaining butter to pan.

Stir milk and cheese into eggs and pour over vegetables. Cook for 3 minutes to set bottom, then bake for 20 to 25 minutes at 350°F until lightly browned.

YIELD: 4 to 6 servings

SLIGHTLY SWEET FRITTATA

An elegant, unusual approach for a special brunch using cooked frozen vegetables. Serve with dates or other dried fruit.

2 cups thinly sliced frozen cooked carrots, partially thawed	¾ teaspoon fennel seeds
	½ cup chopped walnuts
	ground black pepper, to taste
2 teaspoons minced peeled gingerroot	8 eggs
2 tablespoons butter	2 tablespoons water
2 tablespoons honey	1 tablespoon olive oil
¼ teaspoon ground cardamom	1 orange, thinly sliced
¾ teaspoon ground cumin	ground cinnamon

Sauté carrots and gingerroot in butter over medium heat for 1 minute. Stir in honey, cardamom, cumin, fennel seeds and walnuts. Toss to coat carrots with spice mixture and cook for 3 to 5 minutes, or until glazed. Season with pepper. Remove from heat and cool.

In a medium-size bowl, lightly beat eggs and water together. Add three-quarters of the cooled carrot mixture. Reserve remainder for garnish.

Heat olive oil in a 10- to 12-inch ovenproof skillet over medium-high heat. When pan is hot, pour in egg mixture. Cook over medium-high heat, lifting the sides gently so that the uncooked portion flows under the cooked portion. When eggs begin to set on the edges, and the bottom is golden brown (5 to 7 minutes), remove from heat and place under broiler, about 5 to 6 inches away. Broil until top is golden and puffed, 5 to 7 minutes, and frittata feels firm.

Make a circle on top of frittata with remaining carrots. Place orange slices in a decorative pattern around the carrot circle. Sprinkle with cinnamon and serve warm or at room temperature.

YIELD: 4 to 6 servings

VARIATIONS: Use chunks of winter squash or peas in place of the carrots.

* CHEESE, ONION AND *
VEGETABLE QUICHE

This is a good, basic cheese pie recipe, with lots of room for variation. Individual-size tarts freeze very well if frozen slightly underbaked. Can be frozen for 3 months.

1½ cups sliced onions
  2 tablespoons butter
  1 unbaked 9-inch whole
     wheat pie shell (page
     329), or 6 to 8 small tart
     shells, thawed if frozen
1½ cups coarsely chopped
     cooked broccoli,
     cauliflower or
     asparagus, or 1 cup
     chopped lightly
     steamed spinach
     or chard, drained

1½ cups shredded Gruyère,
     Swiss or medium-sharp
     cheddar cheese
  3 eggs
1¼ cups light cream
  ¼ teaspoon paprika
     pinch of ground nutmeg
     pinch of cayenne pepper
     pinch of cayenne pepper

Sauté onions in butter until tender. Cool slightly and spread inside pie shell. Top with chosen vegetable(s) and cheese.

In a medium-size bowl, beat eggs, then beat in cream, paprika, nutmeg and cayenne pepper. Pour over vegetable mixture in pie shell.

Bake in a 350°F oven until top is lightly browned and center is set, 35 to 40 minutes for a 9-inch pie, 20 to 25 minutes for small tarts. Slightly underbake any pies intended to be frozen.

Cool 10 minutes before slicing. Thaw frozen portions before reheating. Bake in 350°F oven for 15 minutes.

YIELD: one 9-inch pie or 6 to 8 small tarts

VARIATIONS: Onion quiche: Increase onions by 1½ cups and omit other vegetables. Top with buttered cracker crumbs before baking.

Herb quiche: Use fresh or frozen herbs in place of the spices. Dill, parsley, chervil, chives and thyme are all good.

Mushroom quiche: Substitute 1 cup mushrooms for the vegetables. Add to the onions while cooking.

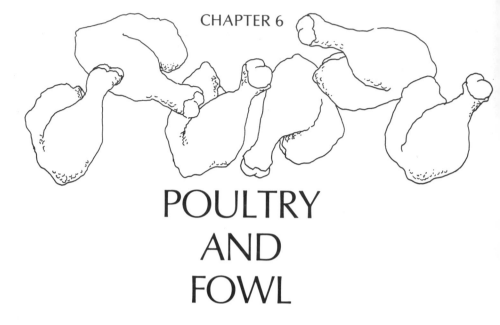

CHAPTER 6

POULTRY
AND
FOWL

If you eat poultry often, you're in the company of the majority of Americans. Almost all of us eat poultry—buttery little marinated broilers, coated and crunchy fryers, big, plump and juicy roasters, turkeys and capons. We eat poultry braised and fricasseed, on its own or smothered with every kind of gravy. We stew it and steam it and poach it and stuff it. We glaze it, we make it Italian or Chinese or Indian by the way we season it, we throw it in salad and turn it to hash, and finally, when every bite is gone, we bet on the wishbone and throw the carcass in the stockpot.

More specifically, 12 billion pounds of chicken pass across our dining tables—domestic and institutional—every year. Why do we love these birds so much? Beyond their versatility, nice taste, low fat and low calorie count, they also excel on strictly utilitarian grounds. Poultry doesn't cost much (adjusting for inflation, chicken is over 40 percent cheaper now than it was fifteen years ago), and it freezes well.

On that latter point there is some disagreement. Tastes vary, after all. Some cooks say frozen chicken is not nearly as nice as fresh, but we say a freezer without poultry is like a Rodale cookbook without whole foods. Freezing changes the texture of poultry and wild fowl less than it does that of meat or fish or vegetables. And this is true of chickens young and old, capons, turkeys young and yearling and Cornish game hens. It's also true of duck, goose, pheasant, quail and squab, all of which are marketed frozen. It's even true of wild fowl that you may rarely have the chance to encounter on the wing, like grouse, guinea fowl, partridge, wild pheasant, pigeon, quail, wild duck and woodcock.

Poultry is a real frozen asset. You can freeze it, whole or in parts. You can defrost it, cook it and freeze it again. It won't suffer noticeably in flavor or texture or nutritional content, especially if you freeze it raw and cook it when it thaws. The only thing you can't do is refreeze poultry that has been frozen and thawed without cooking it first. A whole, raw chicken can be frozen for twelve months, parts for nine months. Cooked chicken, whole or in parts, will freeze for three months. Fried chicken parts can be frozen for four months.

Though poultry prices fluctuate less dramatically than beef prices do, it's still sensible to buy poultry on sale in quantity. The best buys are always whole birds. Parts are a great convenience, but you're paying for someone else to do the simple job of cutting up the frame. You'll save money by doing it yourself (page 118). Only an intrepid one-third of chicken buyers buy chicken whole. The other two-thirds buy cut-up whole chickens, or all-of-a-kind packages, or mixed parts. Of course, if you like wings or dark meat, parts are a boon to your kitchen.

If you decide to buy whole chickens and cut them up yourself, buy big ones. A chicken is a skeleton with meat on it, remember. The smaller the bird, the greater the proportion of bone to meat. Therefore, the bigger the bird you buy, the more meat you are getting for your money. The same goes for turkey. The best way to get value in turkey is to buy it large, frozen and whole.

Don't pass up buying frozen turkey because you think it has to taste inferior to fresh turkey because it's cheaper. Fresh Thanksgiving turkey, if you can find it, has actually gotten to its destination several days before you pick it up. It has endured temperature changes in the plant, the warehouse, the market, your car, your refrigerator and your countertop while you ready it for roasting. Frozen turkeys, on the other hand, are killed and immediately frozen, and they stay that way until you put them in the refrigerator to thaw. A whole, uncooked turkey can be frozen for up to twelve months, parts for six months. Cooked turkey, whole or in parts, will freeze for three months. If a commercially raised turkey is nicely cooked, you really can't tell whether it was fresh or frozen to begin with.

Chicken and turkey are only the highly visible tip of the poultry iceberg. There are many other tasty and unusual forms of fowl that have been domesticated, duck and goose foremost among them.

Unless you have access to a Chinese market or a farm that raises ducks, you'll probably buy ducks that are frozen and go by the name Long Island (because it was Long Island sea captains who first brought them back from China). They usually weigh 4½ to 5 pounds and serve two or three people.

Goose, like duck, is delicious but very fatty. You'll probably most readily find it frozen, and weighing in between 6 and 14 pounds. If you find it fresh at Christmastime, the size will range from 5 to 20 pounds. Both duck and goose have a freezer storage life of only four months because they are so fatty.

If you do manage to come by fresh duck or goose and wish to freeze it, package the giblets separately, because they last only two months in the freezer. Remove the oil glands before freezing, because they can give off a bitter taste before you cook the bird. When you're ready to cook a frozen duck, you will find that it's very wet when defrosted. Dry it thoroughly inside and out with paper towels, or even blow it dry with an electric fan.

Pheasant, quail, Cornish game hen and squab are usually marketed frozen, less often fresh. They're all good roasted or braised. Cornish game hens can stay in the freezer for six to nine months.

Preparing Poultry
for the Freezer

If you buy lean, locally raised, free-range chickens, the initial freezer preparation will fall to you if the seller hasn't done it for you. Start by keeping the chicken in the refrigerator for two or three days to mature and develop its flavor. Refrigeration will also make plucking and drawing easier. Pluck large feathers first, then get at the pinfeathers with a tweezer. Leave singeing the fine hairs for after you've thawed the bird. If you singe it before you freeze it, heat from the flame will start to break down the fat under the skin, making the bird more susceptible to turning rancid.

To draw the bird, make a slit at the vent just big enough to get your hand in. Carefully loosen and remove the intestines, giblets, lungs, kidneys and any fat you find. Make another incision at the back of the neck and take out the crop, windpipe and neck bone. Cut off the feet, and remove the oil sac at the base of the tail. Save the heart, liver, gizzard and neck, and the fat for rendering, but throw out the rest. Handle the fat and the giblets as you would for store-bought poultry. When you've finished, rinse the bird inside and out, scrub down the cutting area and thoroughly wash the utensils you used for cutting.

Preparing Store-Bought Poultry
for the Freezer

Store-bought poultry requires very little cleaning to get it ready for the freezer, so give your attention to choosing it well in the first place. Buy only chicken that is plump, unblemished and odorless, with a soft, cream-colored skin. Prepare it for the freezer as soon as you get it home. Don't put it straight into the freezer in its original wrapper. The plastic is too weak and porous for the freezer. Don't keep it waiting expectantly in the refrigerator, either, where it will stay fresh only a day or two and then be of no use to anyone.

To cut up a chicken, follow the directions given in the text, and do the following:
1. Remove the legs from the body. 2. Separate the drumsticks from the thighs. 3.
Remove the wings from the body. 4. Cut through the skin between breast and
back, splitting the rib cage. 5. Separate breast from back. 6. If desired, divide
breast in half.

Pluck off any pinfeathers, and remove excess fat. If there are giblets, remove and rinse them, and package them separately, because they have a shorter freezer life than the bird does. Rinse the bird, including the cavity, under cold water and pat it dry. Raw poultry can carry salmonella bacteria. Washing before freezing will remove any lurking bacteria that can lie dormant in the freezer but spring to attention as soon as you thaw the bird.

Freeze poultry in the form you intend to use it. If you want to cut it up, do so before you freeze it. Freeze the backs, necks and wing tips separately, and make sure you pack them so they won't puncture the freezer wrap. Save these bony parts for making stock (see recipe for Chicken Stock on page 73). If you freeze a broiler, split it in half down the back and through the breast. Wrap the halves together.

Here's how to cut up a whole chicken into parts for freezing: Begin by placing the chicken breast-side up on a cutting board. Slit the skin at the hip joint, down the bone. Bend the thigh back until the bone pops out of the hip joint. Then cut the thigh away from the body as close to the body as possible. Repeat on the other side. Separate the thigh from the drumstick by cutting through the skin, breaking the joint, and cutting the pieces apart. Repeat on the other leg. To remove the wings, slit the skin at the joint, break the joint, then cut the wing from the body. Repeat on the other side. Cut down through the length of the breast, through the ribs and tail end. This will separate the back of the chicken from the breast. Then cut the breast down the middle to make two pieces. Break the back in half.

If you are freezing a chicken whole, truss it to conserve freezer space and make it oven-ready. To truss poultry, first pull the skin flaps over the cavities of both the neck and tail ends. Fold the wings back and under the bird and tie them

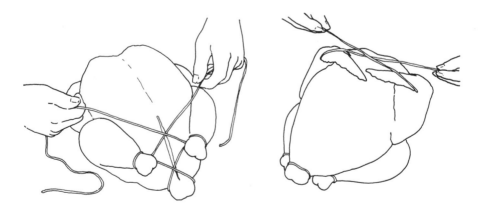

To truss a chicken, lay the bird breast-side up and slip the string underneath its tail. Cross the string ends over the vent, loop each end around a leg, then cross the string over the middle again, pulling the string just enough to draw the legs together. Pull the string back over the breast and down through the wings from the inside out, then up over the body and tie a secure knot.

close to the body with trussing string. When you roast the bird, untie the legs for the last 30 minutes of cooking in order to brown the insides of the legs.

Stuffing a chicken or turkey does not count as making it oven-ready. In fact, quite the opposite is true. Even though poultry and stuffing go together like love and marriage, they don't belong together in the freezer. Stuffings invariably contain ingredients that don't last in the freezer nearly as long as the poultry does. If you don't use the chicken soon enough, the stuffing can spoil and transmit a nasty flavor to an otherwise sound bird. Stuffing also thaws at a different rate than poultry. In case you're wondering why freezing stuffed poultry, so ill-advised at home, is so widespread in the supermarket, it's because commercial processors use special additives and preparation techniques that are not available for domestic use.

Wrapping Poultry

The easiest way to wrap a whole chicken is just to slide it into a freezer bag, neck first, squeeze out as much air as possible with your hands or by immersing the bag in water, fasten the bag with a twist-tie and label it. If you wrap a chicken with freezer wrap of any kind, use the drugstore method of wrapping (page 24) and first pad any protruding parts with foil if they look as if they might break through the wrapping during the rough-and-tumble of freezer existence. If you're wrapping poultry pieces, remind yourself that freezing them all together in a big lump will make thawing a long process. To make it more convenient to use your

To avoid protruding parts of the bird piercing the freezer bag or paper, wrap the ends of the legs and tips of the wings with aluminum foil before placing it in its wrapping.

frozen chicken parts, take the time to wrap each piece individually, then collect all the pieces in a single freezer bag. Label the bag with the number of people it will serve. If you intend to fry or roast the chicken, allow ¾ to 1 pound per serving. If you are going to broil or barbecue pieces, allow half of a chicken or 1 pound per serving. If the chicken will go into a stew, allow ½ to 1 pound per serving.

Preserving the Fat, Giblets and Neck

There's no edible part of a chicken that isn't cozy in the freezer, but fat, as it happens, is just as happy in the refrigerator. If you take the trouble to render chicken fat, it can stay indefinitely in the refrigerator, and be available for basting chicken, mixing into pâtés and sauces, and for sautéing in place of butter, which contains more than twice as much cholesterol as chicken fat. Rendering chicken fat is a simple matter of leaving it in a bowl in a medium oven until it turns liquid. When it cools, strain it into a bowl, and keep it well wrapped in the refrigerator.

Giblets and the neck should definitely be frozen. Rinse and pat dry the gizzard, heart and neck, collect them in a freezer bag and freeze for up to three months. (Uncooked turkey giblets can also be frozen for three months.) Use them to make rich, delicious stocks and gravies. Before you freeze the chicken liver, cut away any green spots in order to save yourself experiencing a bitter mouthful later. Wrap the liver separately from the giblets and neck and freeze it for up to one month (turkey liver will also freeze for a month). Use it, along with any other livers you saved, for pâté or sautéed in a pan for a quick dinner.

Preparing Wild Fowl for the Freezer

Game birds are beginning to appear on restaurant menus, as chefs increasingly pursue uncommon regional American foods for their novelty value and intrinsic good flavor. Special game farms, not accessible to the public except through mail order, are their suppliers. The rest of us have to rely on knowing a hunter or being one. If wild fowl comes your way, you can freeze it without fear, but first clean it without fail. It's very important to clean a game bird as soon as it is killed—not a day or two later, but immediately. The meat, subject to being tainted by the intestines, is at risk from the moment the bird is killed.

To remove the insides quickly, first pluck the bird from the breastbone to the vent. Then slit the skin and the membrane around the vent, up to the breastbone. Put your hand into the breast cavity and up into the neck. When you've loosened the windpipe you can draw out everything inside the bird down to the intestines. (Make sure the intestines and the stomach don't touch the meat. If they do, cut off the meat and throw it away.)

Hang the bird by its head for about two days in a cool place (around 40°F). The length of hanging is a matter of debate among hunters, who recommend times varying from two to five days. Two days is about standard, however. The purpose of hanging is not in dispute—it gives the bird's flesh time to develop in flavor and tenderness.

Pluck the feathers. Pull each one downward in the direction it grows, being careful not to tear the delicate game bird skin. Trim off the wounded area, which would otherwise quickly turn rancid.

Never wash wild fowl meat. Wipe it inside and out with a damp cloth, then treat it as you would poultry, though it will not taste like poultry. It's, well, gamier, and somehow more festive, more of a special treat. Its texture is different, too—it has almost no fat, and its flesh is tougher. Truss it and freeze it whole if it is a young bird, or freeze only the really meaty part of the bird, like the breast of quail, to be sautéed later. If the bird is older, freeze it in sections and then cook it in a casserole or a game pie. The older a bird is, the longer you should cook it. Use the tough, scrawny parts to make delicious stock instead of throwing them out, as many hunters do.

The flesh of wild fowl dries out more quickly than that of domestic fowl, and it should be very carefully wrapped to exclude air. Pack it in a freezer bag and press out as much air as possible with your hands. Or, lower the bag into a bucket of water to release the air. Then fasten the bag tightly. Most wild fowl keeps well in the freezer for about six months. Thaw it in the refrigerator for twelve to sixteen hours, but marinate it as it thaws if you want to tenderize it.

Thawing and Cooking
Poultry and Fowl

Thaw poultry in the refrigerator or let it thaw and marinate simultaneously in the refrigerator. The marinade will solidify at first when it comes in contact with the chicken, but it will liquefy again as the bird thaws.

Whether it's whole or in parts, a chicken should be thoroughly thawed before you cook it, or the interior will not be done when the surface is. You will either have to overcook the outside to get the interior right, or eat the interior undercooked (which makes it subject to the possibility of live bacteria). You can, however, bake chicken parts without defrosting, if you package each one individually. While slow thawing in the refrigerator is safest, you can speed up the process, if you must, by thawing the bird in cold water in a watertight plastic bag. Change the water as it warms up. When the chicken is quite pliable, it's thawed. (Use cold water, even though hot water would seem to make more sense, because hot water would warm the surface too much, increasing the risk of

Uncooked poultry is highly perishable. For bacteria, life begins at 40°F, and poultry should be kept either at temperatures well below that, or above 140°F. Poultry left at room temperature leaves lots of room for bacteria to flourish.

bacteria growth, while the inside is still frozen.) Or, start the bird thawing in the refrigerator, and when it looks as if dinner will never hit the table, finish the job under cold water. It's essential to cook poultry immediately once it's thawed. If you let poultry warm up to room temperature, it will drain juices full of B vitamins, every one of them a thing worth saving.

When you cook poultry that is young or that you've frozen, you may notice some blood in the flesh and bones of the legs. The blood is not an indication of insufficient cooking. What you are seeing is hemoglobin, the edible red pigment in blood, which characteristically leaches out of the bones when a bird that has been frozen is cooked. While it may look a little odd, it won't affect the flavor of the meat, and it's perfectly safe to eat. Returning the bird to the oven will just overcook it.

Here is a summary of thawing times in the refrigerator for poultry and game. A whole chicken weighing less than 4 pounds will take 12 to 16 hours; larger chickens will take 24 to 36 hours. Chicken parts will thaw in 3 to 6 hours; fried parts need 12 hours. Giblets will thaw in 5 hours, as will a pound of livers.

Small turkeys weighing 4 to 12 pounds will thaw in 1 to 2 days, birds of 12 to 20 pounds will take 2 to 3 days and turkeys weighing 20 to 24 pounds need 3 to 4 days. Turkey parts will thaw in 3 to 9 hours, giblets or a pound of turkey livers will take 5 hours.

Allow 24 hours for a Cornish game hen to thaw, 12 to 16 hours for a duck and 24 to 36 hours for a goose.

Freezing Cooked Poultry

If you intend to freeze poultry you've cooked, cool it quickly in a cool place before you package it for the freezer. Cooked dishes that contain poultry will last up to six months in the freezer, but after the third or fourth month they will begin to deteriorate in taste, aroma and texture.

What if you've fried up a mess of crispy golden chicken or roasted a capon perfectly? You can freeze them, but neither fried nor roasted chicken will survive freezing in as tasty or good-looking a condition as it originally was. Poultry does much better in the freezer in combination with other foods or with liquid. You'd be better off taking the meat off the bones of your roasted chicken so you can pack it solidly. Then freeze it with gravy or broth to best prevent contact with air. While slices on their own will last a month in the freezer, they'll last four to six months if they're covered with gravy.

As for stuffing, don't freeze a roasted chicken with the stuffing in it. Freeze the stuffing separately. Nor should you undercook poultry, and then freeze it with the intention of finishing the cooking when you thaw it. You must thoroughly cook it before you freeze it.

Thawing Frozen Cooked Poultry

Frozen poultry can be gradually reheated in the oven or on the stovetop if it's already cooked and it contains a lot of liquid. If you've frozen it whole or in slices, with no accompanying liquid, thaw it in the refrigerator before reheating it. A little hot water running on the freezer container will help loosen the contents so you can transfer them to a saucepan, the top of a double boiler, or a covered pan in a low oven. Stir the contents gently once in a while to prevent sticking. Once you've thawed cooked poultry, it will stay no more than two days in the refrigerator. Don't refreeze it unless you cook it again, in a soup, casserole or other dish, before refreezing it.

BROCCOLI, CHEDDAR
AND CHICKEN PACKETS

Can be frozen unbaked for 4 months if made with fresh chicken. Do not refreeze unbaked packets if made with frozen chicken. Baked packets can be frozen for 2 months.

1 cup broccoli florets
2 whole, boned chicken breasts, thawed if frozen
4 ounces sharp cheddar cheese, thinly sliced

¼ cup Chicken Stock (page 73), thawed if frozen
Lemon-Basil Sauce (page 50)

Steam fresh broccoli until barely tender, frozen broccoli just until hot. Drain well.

Remove skin from chicken breasts and separate into four halves. Make a pocket in each breast half by cutting a lengthwise slit through one side. Set aside.

Tear off four 12 × 12-inch sheets of aluminum foil. Fold in half, then unfold. Lay one sheet of foil on work surface. Arrange one-quarter of the broccoli florets toward center of foil sheet on one half of foil. Top with chicken breast. Stuff one-quarter of the sliced cheese into the packet in the chicken breast. Sprinkle 1 tablespoon of stock over chicken. Seal packet tightly, crimping the edges. Repeat with remaining broccoli, chicken, cheese and stock. If using fresh chicken, the packets can be frozen at this point.

If baking right away, place all four packets on a baking sheet. Bake in 400°F oven for 25 minutes. To serve, open packets, slide ensemble onto plate and top with Lemon-Basil Sauce.

To bake frozen packets, add about 20 to 30 minutes to baking time (carefully open one packet to make sure chicken is done).

YIELD: 4 servings

VARIATION: Replace broccoli with asparagus, about 3 spears per serving.

DELICATE CHICKEN CURRY

This dish will freeze for 6 months.

1 medium-size onion, sliced	½ teaspoon ground ginger
3 pounds chicken parts with bones, partially thawed if frozen	½ teaspoon ground cardamom
	½ teaspoon ground cumin
2 tablespoons vegetable oil	½ teaspoon mace
3 cups water	1 cup sliced carrots, partially thawed if frozen
2 teaspoons turmeric	
2 teaspoons ground coriander	2 cups coarsely chopped celery, partially thawed if frozen
½ teaspoon ground cinnamon	1 cup yogurt, optional

Brown onions and chicken lightly in oil. Add water, turmeric, coriander, cinnamon, ginger, cardamom, cumin and mace.

Cover and simmer gently for 45 minutes. Add carrots and celery. Continue cooking for 30 minutes, or until chicken is cooked through. The chicken can be frozen at this point.

To reheat frozen chicken, thaw in refrigerator, then heat, covered, on top of the stove.

For a more delicate flavor, stir in yogurt just before serving.

Serve with hot brown rice.

YIELD: 4 servings

VARIATION: Add a bouillon cube to the water, or substitute 1 cup of Chicken Stock (page 73) for 1 cup of the water.

PESTO CHICKEN

The pungent pesto and crunchy pine nuts complement this chicken.

2 halved, boned chicken breasts, thawed if frozen	2 cups chopped tomatoes, thawed if frozen
3 tablespoons butter or vegetable oil	12 mushrooms, sliced
½ cup Pesto (page 50), thawed if frozen	¼ cup grated Parmesan cheese
	½ cup chopped pine nuts

Cut four pieces of parchment paper, each large enough to enclose a piece of chicken. Spread each sheet of parchment paper on a work surface and coat the top side of each with butter or oil. Fold each sheet in half, then open it out again. On the right side of each sheet, place a chicken breast and cover it with 1 tablespoon of pesto.

Drain excess water from tomatoes. Arrange mushrooms and tomatoes on top of chicken; cover with remaining pesto. Sprinkle with cheese and nuts.

Form a packet by folding the left half of the paper over the chicken. Fold the top and bottom edges of the parchment paper toward the center to seal them. Then fold the open side over several times to finish the package.

Place on a baking sheet. Bake at 375°F for 18 minutes, or until paper is browned and puffed.

YIELD: 4 servings

CHICKEN LEGS DIABLO

This recipe makes two pans of chicken, one for eating now and one for freezing. If made with fresh chicken, this dish can be frozen up to 4 months. If you use frozen chicken don't refreeze it before baking. The baked chicken can be frozen up to 2 months and reheated later.

8 chicken legs with skin removed, thawed if frozen	3 tablespoons coarse-grained, prepared mustard
½ cup melted butter	1 tablespoon prepared horseradish
5 tablespoons minced chives	

Place chicken in 9 × 9-inch baking pans.

Combine butter, 3 tablespoons chives, mustard and horseradish. Pour over chicken; turn to coat chicken on all sides.

At this point, the pans may be covered and frozen for cooking later if fresh chicken was used. Or, let stand 5 minutes and then bake chicken at 350°F for 50 to 60 minutes, or until flesh is tender and juices run clear when thigh is pierced through its thickest part. Bake frozen Chicken Legs Diablo covered for 60 to 90 minutes, or until tender.

Serve on a heated platter, sprinkled with remaining 2 tablespoons chives.

YIELD: two 4-serving casseroles

HONEYED CHICKEN STIR-FRY

4 boned chicken breasts, partially thawed if frozen	1 cup broccoli florets, partially thawed if frozen
2 tablespoons honey	1 cup sliced carrots, partially thawed if frozen
2 tablespoons soy sauce	
2 cloves garlic, minced	
1 cup chopped celery, partially thawed if frozen	2 tablespoons vegetable oil
	4 scallions, including green tops, chopped

Skin the chicken breasts and cut into ½ × 1½-inch strips.

In a bowl, combine honey, soy sauce and garlic. Add chicken and toss to coat. Let stand 20 minutes.

Meanwhile, if using fresh celery, broccoli and carrots, steam them for 3 minutes and reserve.

Heat oil in wok or skillet over medium-high heat. Add chicken and sauce. Stir-fry for 3 to 4 minutes, or until chicken is white. Remove with a slotted spoon and keep warm.

Add scallions, celery, broccoli and carrots to wok. Stir-fry for 3 to 4 minutes, until crisp-tender. Return chicken to pan and stir to combine.

Serve over hot brown rice.

YIELD: 4 servings

VARIATION: Substitute 8 to 10 cherry tomatoes, halved, for the carrots. Stir in immediately before serving.

CHICKEN STIR-FRY
WITH SNAP PEAS

2 teaspoons Mochiko rice
flour
¾ cup cold water
¾ teaspoon powdered soup
concentrate
1½ tablespoons soy sauce
¼ teaspoon dried tarragon
4 boned chicken breasts,
partially thawed
if frozen
3 tablespoons vegetable oil

1 small onion, sliced
1 clove garlic, minced
¾ cup sliced sweet red
peppers, partially
thawed if frozen
½ cup sliced water
chestnuts
1 cup snap peas, thawed
if frozen
4 scallions, including tops,
chopped

In a small bowl, mix rice flour with ¼ cup of the water. Stir until smooth.
Add remaining ½ cup water, soup concentrate, soy sauce and tarragon.
Set aside.

Slice skinned chicken into ¾-inch cubes. Heat 2 tablespoons of the
oil in a wok. Add chicken. Cook and stir until meat turns white, 4 to 5
minutes. Remove chicken from wok with a slotted spoon and keep warm.

Add remaining oil to wok. Add onions, garlic and red peppers. Cook,
stirring constantly, just until crisp-tender, 2 to 3 minutes.

Add water chestnuts, snap peas and rice flour mixture. Cook and
stir until sauce is thickened and clear. Stir in chicken and scallions and
cook 5 minutes, stirring occasionally.

Serve with hot brown rice.

YIELD: 4 servings

VARIATION: Use 1½ pounds frozen shrimp, shelled and deveined, in
place of the chicken. Thaw the shrimp just enough to separate. Omit
soy sauce or decrease to 1 tablespoon. Decrease cooking time for shrimp
to 3 minutes.

GINGER CHICKEN STIR-FRY

2-3 boned chicken breasts,
    partially thawed
    if frozen

3 tablespoons vegetable oil

1 clove garlic, crushed

1 teaspoon grated fresh
    gingerroot

1 stalk celery, diced

2 scallions, including
    green tops, thinly
    sliced

1 medium-size sweet red
    pepper, diced

8 ounces snow peas,
    partially thawed if
    frozen

½ cup sliced water
    chestnuts

¼ cup coarsely chopped
    cashews

½ cup water

2 teaspoons Mochiko rice
    flour

2 teaspoons soy sauce

Slice skinned chicken into 1 × 2-inch pieces. In a wok or Dutch oven, heat 2 tablespoons of the oil over medium heat until hot. Add garlic and sauté for 2 minutes. Remove and discard garlic.

Add chicken and stir-fry for 5 minutes, or until white. Remove with a slotted spoon and keep warm.

Add remaining tablespoon of oil to the wok. Stir-fry the gingerroot, celery, scallions and peppers for 2 minutes. Return chicken to wok. Add snow peas, water chestnuts and cashews. Stir-fry for 3 minutes.

In a cup, combine 1 tablespoon water and the rice flour. Stir. Add soy sauce and remaining water. If using frozen snow peas, omit ¼ cup of water. Add flour mixture to wok. Stir for a minute or two, until sauce thickens slightly.

Serve over hot pasta or brown rice.

YIELD: 4 to 6 servings

CHICKEN AND SAUSAGE CASSEROLE

Can be frozen for 6 months.

4 chicken breasts, partially
  thawed if frozen
1 stalk celery, chopped
2 bay leaves
¼ teaspoon ground black
  pepper
1½ cups sliced carrots
1 pound bulk hot sausage
1 pound mushrooms,
  sliced, about 4 cups

3 cups Basic Tomato Sauce
  (page 56), thawed
  if frozen
2 teaspoons soy sauce
2 teaspoons Worcestershire
  sauce
5 cups cooked brown rice
1 cup cooked wild rice

Place chicken in a large pot and cover with water. Add celery, bay leaves, pepper and carrots. Simmer until chicken is tender, about 1½ hours. Remove chicken. Discard bay leaves and reserve ½ cup of the stock and the celery and carrots. The remainder of the stock can be frozen for future use in other recipes. Bone and skin chicken and cut chicken into 1-inch cubes.

In a skillet, cook sausage until brown. Break into small pieces and pour off fat. Add mushrooms to the skillet and simmer until soft. Add 1½ cups tomato sauce, soy sauce, Worcestershire sauce, and reserved stock, celery and carrots.

In a large bowl, add remaining tomato sauce to brown and wild rices and stir. Layer the chicken, rice mixture and then the sausage mixture into two buttered 2-quart casserole dishes. Freeze one for later use. Bake the other uncovered at 375°F for 30 minutes.

To make the frozen casserole, bake, covered, at 375°F for 1½ hours, uncover and continue to bake until hot and bubbly (about 20 minutes).

YIELD: 12 to 14 servings

TERRIFIC TURKEY TRIANGLES

Freeze your leftover cooked turkey for use later as an appetizer or a main course. The triangles can be frozen for 3 months.

4 cups chopped cooked
   turkey, thawed if
   frozen
8 scallions, including
   green tops, thinly
   sliced
1 teaspoon ground black
   pepper

20 ounces cream cheese,
   at room temperature
¼ cup minced fresh parsley
24 phyllo leaves
1 cup melted butter,
   approximately

In a medium-size bowl, combine turkey, scallions, pepper, cheese and parsley.

Place a phyllo leaf on a flat surface. Brush with melted butter. Repeat until you have a stack four leaves high. With a sharp knife, cut into four strips across the short side.

Place about 2 tablespoons of turkey mixture onto the end of one strip. Fold the strip, starting at end with turkey mixture, into a triangle shape. Brush top with melted butter. Repeat with the other three strips.

Repeat the entire procedure until you have used all the phyllo.

At this point you can freeze the triangles for up to 3 months or bake immediately at 350°F for 25 minutes.

When ready to bake frozen triangles, place on a cookie sheet. Bake at 350°F for about 35 minutes, until lightly browned.

Cut in half for appetizers. Serve hot.

YIELD: 24 triangles (48 servings as appetizers, 12 servings as main course)

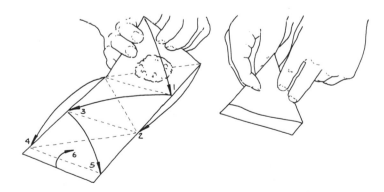

Begin folding the turnover by picking up the lower righthand corner closest to you and folding it over the turkey mixture to the left, forming a triangle with the first fold. Continue folding as shown by the dotted lines and arrows. (It's like folding a flag!)

TURKEY PIE

Here's a good way to use leftovers. Prepare and freeze this dish in advance for a winter supper or buffet. Can be frozen without potato crust for 6 months.

1 green pepper, minced	½ cup butter
1 cup sliced carrots, partially thawed if frozen	¼ cup whole wheat flour
	2 cups turkey stock or Chicken Stock (page 73), thawed if frozen
⅔ cup chopped celery, partially thawed if frozen	1 cup half-and-half
2 onions, minced	¼ teaspoon ground nutmeg
10–12 mushrooms, sliced	pinch of cayenne pepper
3 tablespoons vegetable oil	4 cups cubed cooked turkey, thawed if frozen
2 cups peas, partially thawed if frozen	
½ teaspoon dried sage	6 potatoes
1 teaspoon dried thyme	½–¾ cup milk
1 teaspoon dried basil	¼ cup grated Parmesan cheese

Sauté green peppers, carrots, celery, onions and mushrooms in oil until limp. Blanch peas in simmering water for 3 minutes, then drain and add to other vegetables. Stir in sage, thyme and basil.

Melt 6 tablespoons of the butter in a medium-size saucepan and cook until foamy. Stir in flour and cook a minute or two over low heat. Whisk in the stock and continue cooking, stirring constantly, until mixture begins to thicken. Stir in half-and-half and cook until thick enough to coat the back of a spoon. Add nutmeg and cayenne.

Combine sauce with cooked vegetables and turkey and pour into oiled 9 × 9-inch baking dish or two 9 × 5-inch pans. At this point, the casseroles may be covered and frozen for later use. To finish the preparation prior to serving, partially thaw the turkey mixture in the refrigerator, top with potato crust and bake at 350°F as directed below. Baking time will increase about 20 minutes.

To make the crust, cook potatoes in simmering water until tender. When cool enough to handle, peel and mash the potatoes. Beat in milk, remaining 2 tablespoons of the butter and 2 tablespoons of the cheese. Spread potato mixture over top of turkey mixture. Sprinkle with remaining cheese. Refrigerate until ready to bake. Bake at 350°F until warm and bubbly and lightly browned on top, about 1 hour.

YIELD: two 4-serving casseroles

VARIATION: Substitute leftover gravy for the butter, flour and stock. Just reheat it before adding the half-and-half.

GLAZED CORNISH GAME HENS

2 Cornish game hens,
  thawed if frozen
¼ cup chopped onions
½ cup diced Golden
  Delicious apples
  freshly ground black
  pepper, to taste

2 tablespoons butter
3 tablespoons red wine
  vinegar
3 tablespoons currant
  jelly

Place hens in a medium-size roaster.

In a small bowl, combine onions, apples and pepper. Place ½ tablespoon of the butter in cavity of each hen. Stuff each hen with half the onion-apple mixture. Tie the legs together with string.

Rub the outside of the hens with remaining butter. Sprinkle with additional pepper.

Roast at 400°F for 20 minutes; reduce heat to 350°F and roast for an additional 30 to 45 minutes, until juices run clear and legs move freely. Brush hens with pan drippings every 10 minutes.

Remove pan from oven and let stand for 5 minutes.

Place hens on a serving platter. Remove strings and discard.

Place the roasting pan over medium heat (probably using two burners). Scrape up the brown pieces on the bottom and stir in the vinegar and jelly. Cook the glaze, stirring constantly, until syrupy, 5 to 7 minutes.

Brush glaze over hens several times and serve.

YIELD: 2 servings

LEMON CORNISH HENS
STUFFED WITH BULGUR AND FRUIT

Bulgur is the traditional grain of the Middle East, where it's often combined with fruits. This recipe melds the Middle Eastern tradition with an Oriental one—in this case, lemon-marinated Cornish game hens. The stuffing can be frozen by itself for 4 months.

2 onions, minced	¼ cup raisins
¾ cup lemon juice, thawed if frozen	½ lemon, thinly sliced
4 Cornish game hens, thawed if frozen (reserve livers from hens and chop)	2¾ cups Chicken Stock (page 73), thawed if frozen
½ cup chopped almonds	3 tablespoons chopped fresh parsley
3 tablespoons butter	1 teaspoon mint leaves
2 cups bulgur	vegetable oil
¼ cup chopped dried apricots	

In a large bowl, combine onions and lemon juice. Marinate hens in this mixture for at least 20 minutes at room temperature. Turn hens several times so that they marinate evenly. Drain off the marinade and reserve. Place hens in a roasting pan.

Sauté livers, almonds and onions in butter for about 5 minutes. Add bulgur and continue cooking about 3 minutes, until bulgur is lightly toasted. Set aside.

Simmer apricots, raisins and lemon slices in 1¾ cups of the stock for 3 minutes. Add to bulgur mixture and simmer for 5 minutes. Stir in parsley and mint. At this point, the stuffing may be cooled and frozen for later use.

Stuff each hen loosely with the bulgur mixture. Extra stuffing can be heated separately in a baking dish. Rub each hen lightly with oil. Add remaining 1 cup stock and marinade to the pan and bake at 350°F for 1 hour and 15 minutes. As the hens cook, baste several times with the pan drippings.

YIELD: 4 servings

ROAST DUCK AND ORANGE SALAD

1 medium-size green
  pepper
3 navel oranges
2 cups ¼-inch julienne
  strips roast duck,
  thawed if
  frozen
1 teaspoon chopped fresh
  gingerroot
1 small clove garlic,
  minced
¼ cup no-salt peanut butter

2 tablespoons rice wine
  vinegar
¼ teaspoon chili oil
¼ cup peanut oil
1 tablespoon orange peel,
  minced
½ cup orange juice
  freshly ground black
  pepper, to taste
2 heads Boston lettuce
1 bunch arugula or red
  lettuce

Slice green pepper lengthwise into ⅛-inch slices. Peel and section oranges, removing the membranes that surround the individual sections. Place peppers, oranges and duck in a medium-size bowl.

Combine gingerroot, garlic, peanut butter, vinegar, chili oil, peanut oil, orange peel and juice and pepper in a food processor or blender and blend until thoroughly mixed. Pour over peppers, oranges and duck and toss to coat all the pieces.

Wash and tear the lettuce into serving-size pieces. Wash the arugula and remove any large stems. Line a serving bowl with the lettuce and arugula and spoon in the duck mixture.

YIELD: 4 servings

STUFFED DUCK WITH PRUNE SAUCE

Duck

½ cup chopped onions
½ cup chopped celery
1 clove garlic, minced
3 tablespoons butter
2 tablespoons chopped fresh parsley
½ teaspoon soy sauce
⅛ teaspoon ground black pepper

½ teaspoon dried thyme
3 cups cubed whole wheat bread
¼ cup Chicken Stock (page 73), thawed if frozen, or water
1 frozen 4- to 5-pound duckling, thawed

Prune Sauce

1 cup chopped prunes
1 teaspoon grated orange peel
2 teaspoons white wine vinegar

2 tablespoons honey
2 cups water

To prepare the duck, in a medium-size skillet over medium heat sauté onions, celery and garlic in butter until onions are translucent, 5 to 7 minutes. Add parsley, soy sauce, pepper and thyme to onion mixture and mix well. Toss with the bread cubes in a medium-size bowl. Add chicken stock and toss well.

Spoon stuffing into duck and secure opening. With a sharp fork, prick duck in several places. Place in a baking pan on a rack. Bake at 350°F for about 2 hours, or until tender, turning duck over several times so fat can drain off.

To make the sauce, combine prunes, orange peel, vinegar, honey and water in a medium-size saucepan. Bring to boil over medium heat. Reduce heat to low. Cover and simmer for 30 minutes. Remove from heat. Rinse blender bowl in hot water. Puree prune mixture in blender until slightly chunky. Makes about 1½ cups of sauce. Serve hot sauce over duck.

YIELD: 3 to 4 servings

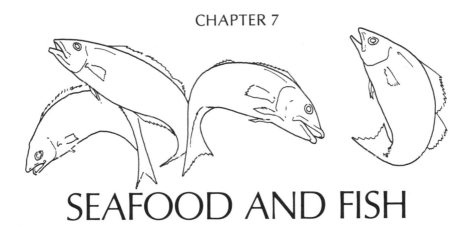

CHAPTER 7

SEAFOOD AND FISH

Freshly caught fish are such a treat they deserve your most considerate handling. They taste their very best when you've just pulled them out of the water, cleaned them and aimed them straight for the frying pan. So when your catch is substantial, sit down to enjoy as much of it as you can while it's fresh. Even if you keep the remainder tightly wrapped in the refrigerator, the very longest it will stay safe to eat is three days.

The sooner you get surplus fish into the freezer, the closer to fresh-caught its flavor will be when you cook it. Once fish has been frozen, it won't taste exactly the same as when it was freshly caught—there's no fish like a fresh fish—but it will be tender and moist and free of fishy odors. Fish is almost 20 percent protein, and it will retain almost all of its hefty nutritional value in the freezer. To get the maximum benefits from frozen fish, however, you'll have to pay heed to several fish foibles while you prepare it for the freezer.

Keeping Fish Cold

A cold fish is a happy fish. Commercial fishermen always either flash-freeze their catch or hold it on ice. Their way should be your way. Keep your catch cold from the moment it comes off the hook. Ideally, each fish should go straight into an ice chest that you've packed with bags of ice cubes or plastic freezer packs. But that's not always possible. When it isn't, which is probably most of the time, try to keep the fish alive as long as possible, in the water if it is cold. Then gut it at streamside, and wipe out the inside. At the very least, try to keep the fish in the shade until you can get it into a skillet or a freezer. If you're lucky enough to catch a lot of fish, pack twigs or boughs between them when you get them ready to take home, to keep air circulating. There's a good reason for all this caution: Fish is the most perishable of all foods.

Fish have the most efficient digestive system known. As soon as a fish dies, the enzymes in its digestive tract go to work breaking down the fish's own flesh, since they have no other food to break down. For these hard-working enzymes, fish flesh, which is highly digestible to begin with, is child's play.

When you want to store the fish you've caught, not only do you have to race against the fish's amazingly aggressive digestive system, but you must also contend with aggressive bacteria both inside and outside the fish. Normal refrigerator temperatures do not long stave off these cold-loving bacteria because even below 32°F they are active, when most other bacteria are not.

The high degree of unsaturated fatty acids that exist in fish also turn it rancid more quickly than happens in meat where the ratio of unsaturated to saturated fatty acids is lower. Not even freezing can keep fish safe as long as meat. It can't be overstressed that with fish it's essential that you move quickly to chill it, clean it, eat it or freeze it. Just about the worst thing you can do is to keep a whole, dead fish around unrefrigerated. Small fish spoil faster than large ones and freshwater fish spoil faster than saltwater fish, but they all spoil sooner than any other food.

Traveling Long Distance with Fish

If you must drive a day or more to get home, fish will make suitable traveling companions only if they are frozen. Otherwise, their presence can be a real hindrance to good conversation. Try this method if you have access to a freezer before you start for home: Clean the fish immediately. Lay them out in big shallow pans, cover them with water and freeze. When you're ready to leave for home, dip the pans in water to free the ice blocks, then pack the blocks in a picnic cooler. Fill in any gaps with shredded newspapers. Tape all around the edges of the lid to seal out air, and the blocks should stay frozen for up to two days in warm weather, longer in cold weather. When you get home, wrap the blocks in freezer wrap and put them in the freezer.

What if you've got access to freezing facilities but no shallow pans? Bring along or buy some heavy-duty aluminum foil and make your own shallow freezing trays. Rip off a sheet, double over the edges and bend them up to form sides. Add a fish, some water to cover, and freeze.

Cleaning and Dressing Fish

When you clean a fish you get to know it very well. The idea that cleaning a fish is difficult or distasteful is wrong. When it is very fresh, as it should be when you clean it, fish smells only of the water and its environs. It's redolent of the place where you had such a good time catching it.

A fish that has been cleaned is referred to as dressed. Having lost several body parts, though, it is actually more undressed than it was before. Dressed means that for cooking and serving purposes the fish is in respectable shape.

There are many ways to dress a fish, and your choice will depend on what kind of fish it is and what you'll eventually do with it. Before you start to work on it, think how you will use it when it is thawed. Will it end up as chunks in soup, filleted and poached as a cold first course, stuffed and baked whole as a main course? Your goal when you dress the fish is to prepare it so it will be ready to cook with a minimum of handling as soon as it is thawed. Fish starts to get mushy quickly when it thaws.

Removing Scales and Fins

If your catch is trout, salmon, haddock or striped bass, consider serving it whole. Remove the scales and fins and gut it by the method described below, and you will have a "whole-dressed" fish. A "pan-dressed" fish is prepared the same way, but the head and tail are also removed to make it fit better in the pan. "Halved fish" are pan-dressed fish that are cut in half along the backbone. This is the best form for grilling and broiling because the skin and bones remain to hold in moisture when the fish is exposed to intense heat.

Start dressing the fish by removing several parts that will never be missed, beginning with the pectoral and the pelvic fins. These should go first because they may be sharp. Cut them off with kitchen shears or a sharp knife.

The scales are also removable, except in the case of catfish and eel, which have a leathery skin that must be completely removed, and trout, whose scales are integral to the skin and are not removed. If you are working indoors, spread lots of newspapers around to catch the scales as they fall, or scale the fish underwater to keep the scales from flying around. Hold the fish firmly by the tail and run a scaler or the dull side of a knife blade from tail to head (away from your body), repeating the motion until you can't see any more scales.

That done, remove the dorsal and anal fins. Cut the flesh on either side of them so you can pull them out. If you're going to cook the fish whole, the dorsal and anal fins can stay where they are. If you are going to eat the fish within 24 hours you can also leave them in place. If you're going to freeze the fish or keep it in the refrigerator longer than one day, cut off the gills, too. They are even more perishable than the flesh and start to smell bad even faster.

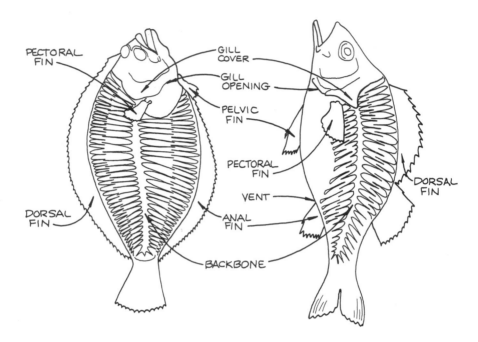

Skeletal structure of the flatfish, left, and the roundfish, right.

How you remove the innards depends upon whether you're confronting a roundfish or a flatfish. If it's a roundfish, slit the belly from the anal opening to the gills. Save the roe in their long sacs, if you come upon any, to cook or freeze separately. Pull out the internal organs, then run your knife along both sides of the backbone to release any remaining pockets of blood. Blood that is not washed away can discolor the flesh while the fish is in the freezer, so rinse the fish quite thoroughly, then rinse it again.

If you are gutting a flatfish, you can reach the innards through the gill opening since they are all neatly collected right there near the head. Hook your thumbs in the gill opening and pull it open as far as you can without tearing the skin and flesh. Reach right in and pull out the organs. Check that everything that was inside is now outside, then rinse the fish until it is thoroughly clean of blood and bits.

If you don't like the idea of looking a fish in the eye, remove its head. Consider, though, that retaining the head helps prevent moisture from leaking out during cooking. It also provides a little drama at the table to serve a fish whole, perfectly broiled or steamed and afloat in a complementary sauce. The same is true of the tail. Retaining the head and tail enriches the flavor of the fish as it cooks, and of the sauce it cooks in. It also keeps the fish warm longer when you serve it.

If you decide it's off with the fish's head, cut it off just behind the gills with a sharp knife if the fish is small or with a knife and a cleaver if the fish is large. With small fish you can first snap the backbone in your hand or against the side of a table.

The heads, tails and skeletons of the fish you freeze should find their way into a plastic bag fitted into a plastic container. Freeze the parts and save them to use later in making your own delicious fish stock (page 78) and sauces.

Cutting Steaks

Steaks are a convenient form for freezing large firm-fleshed fish for later broiling or sautéing. Halibut, salmon, cod, snapper, sablefish and swordfish are all good candidates. To cut a fish into steaks, remove the scales and fins, the innards, head and tail. Then with a sharp knife cut the fish in 1½-inch-thick slices across its width. Start each slice by cutting down through the backbone while the fish is on its belly. Then turn it on its side and cut through the entire body.

If you want to remove the skin, do it after you cut the slices. Just slip a sharp knife between the skin and flesh at one end while you grip the slice at the other end. Slice back and forth gently, pulling the skin away with your other hand.

To prepare chunks, cut steaks, then divide the steaks into pieces, discarding the backbone. Or you can cut fillets into chunks. Use the chunks in fish stews and casseroles.

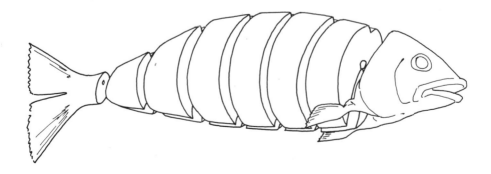

To cut fish steaks, begin at the head end. Hold the fish belly-side down and use a large knife to cut down through the flesh and backbone. Then turn the fish on its side and cut through the remaining flesh. Slices should be about 1 to 1½ inches thick, according to the size fish.

To fillet a roundfish, hold the fish on its side, tail end toward you and make a cut from the head end to the tail along the backbone, left. Then make a cut behind the gills down to the backbone, center. Hold the knife parallel to the ribs and make short strokes toward you, separating the flesh from the bones, right. Gently lift the flesh with your other hand as you cut. Continue over the entire side until the fillet lifts free. Turn the fish over and continue for the second fillet.

Cutting Fillets

Take away its skin and bones, a fish becomes a fillet. It is fragile in this form and less tasty without the bones to enhance its flavor, but it *is* convenient. Fillets are the most popular cut of fish commercially available because of the absence of bones. The large, flat fishes of the flounder family, in particular, lend themselves readily to filleting, although many kinds of fish can be filleted.

Here's how to fillet a roundfish. Begin by gutting and washing the fish. Hold the fish on its side with the tail toward you. Make a slice along the backbone from the head to the tail deep enough for you to actually see the backbone. Sever the fillet at the head by cutting down to the backbone behind the head. Holding the knife parallel to the ribs and using short strokes, continue to cut down toward the tail between the ribs and the flesh, pulling the fillet up with the other hand.

Use the bones to guide your knife, and continue cutting along the entire length of the fish.

To cut the fillet from the other side, hold the fish by the backbone and use your knife to cut the flesh from the ribs. Some of the bones will remain in fish with fragile skeletons. Pick them out with your fingers or tweezers.

To skin the fillets, lay the fillet on a board, skin-side down. Slice into the fillet very carefully between the flesh and the skin at the tail end. Hold the fish by the piece of loose skin. With the knife held at a shallow angle and using short strokes, cut from the tail to the head until the entire fillet is separated from the skin. Always cut away from yourself.

If you are filleting a flatfish, you will wind up with four fillets, two on each side. First lay the fish down on a board, tail toward you. With a sharp knife, cut down to the backbone along the center of the fish from the head to the tail. Holding the knife at a shallow angle and beginning at the head, use short strokes to sever the flesh clear of the ribs. Continue cutting along the entire length of the fish, letting the bone structure guide your knife. Cut the fillet off at the tail end. Trim any ragged edges. Use this method to cut off the other fillet, then turn

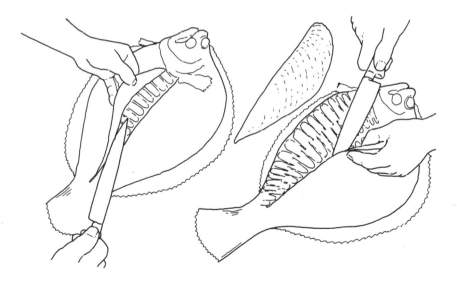

To fillet a flatfish, lay the fish on its side and gently make a cut down the backbone in the center of the fish, left. Hold the knife at a slight angle, almost parallel to the fish and use short strokes to separate the flesh from the bones. Lift the flesh up with your other hand while cutting until the entire fillet is separated from the fish. Repeat on the other half of the fish, right, then turn the fish over and remove the other two fillets.

the fish over and repeat the procedure until you've cut four fillets. Skin the fillets as you would roundfish fillets.

Freezing Fish the Fisherman's Way

Fishermen trust their fishing instincts. Those instincts differ and so, too, do their methods for freezing the fish they catch. That's why you're about to read about several different ways to freeze fish. Tests at the Rodale Test Kitchen were inconclusive as to which method was best, but all of them worked.

Freezing Fish in an Ice Block

Most fishermen agree that freezing fish in water creates the best airtight covering. Here's the fastest way to do it. First clean and dress the fish. Partially fill a freezer container with water, or even better, a cardboard milk carton large enough to hold the fish just surrounded by water. Add the dressed fish (it doesn't matter whether the fish is whole, in chunks, steaks or fillets). Fill the container with enough cold water to cover the fish, but leave an inch of headspace for water expansion. Shake the container gently to make sure the water completely surrounds the fish. Run a knife around the sides of the container to burst any air bubbles. Put the lid on the container, or seal the carton with freezer tape, then label and freeze. That's all there is to it. This method works for both saltwater and freshwater fish. A variation of this technique, freezing fish in a panful of water, was described on page 140.

Ice Glazing Fish

Ice glazing works on the same theory as the method described above, namely, the wetter a fish is when you freeze it, the better the results will be. To form the protective ice layer between the fish and the air, first tray-freeze the fish until it is solid, about two or three hours. (If your fish is as fresh as it should be, it will be odorless and safe to keep unwrapped in the freezer.) Dip the frozen fish in cold water, place it in a freezer bag and return it to the freezer. This procedure should be repeated a number of times, as the thicker the ice glazing, the more protection for the fish. The virtue of the ice glaze method is that it allows you to gather several frozen fish together in one bag, and each will remain separate. For fillets, which are especially hard to separate, freezing and then ice glazing is a particularly handy way to keep them apart in the freezer. They can be cooked evenly while still frozen, and they'll be juicy and flavorful.

Dipping Fish

Fatty fish has a freezer life of only two to three months, but there are a couple of ways you can prolong that time and still keep the fish's flavor good and its flesh firm. Try dipping the fish either quickly into lemon juice, or for twenty seconds in a bath of 2 teaspoons ascorbic acid crystals (vitamin C, available in drugstores) dissolved in 1 quart of cold water. While the fish is still wet, wrap it and freeze it. Since all fish roe is somewhat fatty, use either of the dipping methods above before freezing it, no matter what the characteristics are of the fish it once called home.

Tray-Freezing Fish

If you want to be truly simple about freezing fish, you can spread them out on a cookie sheet, and when they are frozen solid, wrap each one individually and return it to the freezer. No dunking, no dipping. An even simpler way is to wrap the fish and freeze it. This is the quickest and easiest way to freeze fish. Unfortunately this method has mainly speed to recommend it. The texture of fish frozen this way, lean fish in particular, like haddock or cod, can get mushy.

Wrapping and Labeling Fish

You already know by now that you have to wrap food very tightly to store it in the freezer. Fish is certainly no exception. Whatever freezing method you use, make sure that the fish is completely covered and that the covering is securely fastened. When you freeze several pieces of fish together, unless you have tray-frozen the fish first and then ice glazed it, the pieces will surely stick together. Wrap each piece individually to avoid ending up with a frozen clump of fish.

Label each package with the kind of fish it contains, the quantity, the date frozen and even the maximum length of storage. Note where you caught the fish, if you are likely to forget, and when you take it out of the freezer to thaw you'll be able to reminisce a little.

How Long Can You Keep Fish in the Freezer?

There's no easy answer to that question. It very much depends on the kind of treatment the fish received before it got to the freezer as well as the kind of fish you're freezing. If it is a fatty fish, two to three months is the maximum storage time, but you can extend its tenure to nine months if you cut away the loose flaps just below the ribs (these are often particularly evident in saltwater fish). Otherwise, the oil in the flaps will penetrate the rest of the fish and give it an off-taste after a few months. Fat just doesn't freeze as well as carbohydrates or proteins.

Lean fish has a freezer life of four to six months in peak form. If you dip fish, whether it's lean or fatty, in lemon juice or a solution of ascorbic acid crystals in water, you can extend its freezer life by about three months.

LEAN AND FATTY FISH

LEAN FISH	pickerel	carp
	pike	catfish
black drum	pollack	eel
blackfish (tautog)	red snapper	hake (whiting)
black sea bass	rockfish	herring
blowfish tails	scrod	kingfish
(sea squab)	shark	mackerel
cod	skate	mullet
croaker	squid	pike
cusk	tilefish	salmon
flounder	weakfish (sea trout)	shad
gray sole	yellow perch	smelts
haddock	yellow pike	swordfish
halibut		trout (freshwater)
lemon sole	FATTY FISH	tuna
monkfish		whitefish
ocean perch	bluefish	
	butterfish	

Thawing Fish

Thaw fish overnight in the refrigerator, where it will take six to eight hours per pound. If you forgot to take your fillets out of the freezer ahead of time, you can cook them frozen; just double the cooking time. Whole fish, on the other hand, is usually too thick to cook straight from the freezer. If you cook frozen fillets, you won't be able to bread or stuff them, of course, because the breading won't adhere, and the fish will be too inflexible to stuff. And if you broil them, move the broiling rack farther away from the heat source than you would normally have it. Frozen fish takes twelve to fifteen minutes per inch to cook.

If you forgot to move your fish from freezer to refrigerator a day ahead of time, you can thaw it quickly under running water. Put the fish, still wrapped, in a deep bowl, and let a gentle stream of cold tap water run over it. Thawing fish in cold water is a quick method that gives excellent results. A pound of fish will thaw this way in an hour or two at most.

If you froze your fish in an ice block, rip open the wax carton, running the carton under warm water very briefly to loosen it. Then thaw the ice block under cold running water.

When is fish thawed? As soon as it is pliable. An odd ice crystal or two won't matter at all. The trick with thawing fish is to keep it cool until you cook it. Once it's thawed, treat it just like fresh fish. It will deteriorate rapidly, so cook it immediately.

The Fish You Shouldn't Freeze

Don't refreeze fish that you've already frozen once. It's not that resilient. And don't freeze fish that you've cooked. Reheating it will take as long as the original cooking, and you'll end up with fish that's been cooked twice as long as it should have been. The flavor will be ruined and the nutrients all but destroyed. The exception to this rule, and of course there is an exception, is this: You can freeze fish in pies, fish cakes or in a sauce. Slow reheating in combination with other foods will help the fish to retain its moisture and its flavor.

Another bit of advice: Don't freeze the "fresh" fish you buy from a supermarket or even a fish store unless you know you can trust them. Even though you're buying it "fresh," it may well have been frozen once either on the boat or by the wholesaler. Your fresh fish may just be fresh from frozen storage.

Cooking Frozen Fish

Fish is so wonderful to eat that the best thing you can do when you cook it is to leave it alone. Simplicity is best where fish is concerned. Don't assume that the fish you froze is so inferior that you must go to great lengths to make it presentable. You don't have to marinate it or sauce it or stuff it unless you want to. Frozen fish, properly prepared for freezing and then thawed just to the slightly icy stage, is good enough that you can serve it simply broiled or baked as you would fresh fish.

We divide fish into many categories: freshwater and saltwater, flat and round, and for cooking and freezing purposes, lean and fatty. This last category is useful but not entirely accurate. Some fish fall somewhere between lean and fatty, and some fish are lean in one locale, season and level of maturity, and fatty in another. Despite the variations, however, the classification has some validity. Lean fish like haddock and cod are less than 1 percent fat. Fatty fish like salmon and mackerel are more than 14 percent fat. You may well want to treat lean and fatty fish differently in the kitchen. Lean fish is generally white fleshed and delicate. It responds admirably to being baked in a sauce, or basted often but lightly with butter if you broil it. Lean fish benefits from any help you can give it to retain its

---

If you cook fish too long, the tissues break down and provide an easy escape route for flavor and juices. Cook fish only until it flakes easily; any longer, and your fish will become tasteless and dry.

---

moisture, but don't overdo it—light is best. A little embellishing with onions, garlic, tomatoes, thyme or dill is a good idea. Don't overcook it. It's already tender enough.

Fatty fish broils and bakes easily because it contains its own oils in such abundance that it needs very little outside help to develop its rich flavor. If that flavor is, in fact, a bit too much for you, you can cut its strength by basting it with citrus juice, thin tomato juice or vinegar, again lightly. A fatty-type fish that you season lightly, wrap in foil and barbecue makes a memorable treat on a summer evening.

Shellfish

Clams

It's hard to imagine anyone digging clams and not consuming them within the hour, but if you want to freeze them for a later time, it will work. Discard any clams that are not tightly closed or that do not close immediately when you tap them. Closing its shell is the primary defense mechanism of a clam. If it doesn't use it when frightened, the creature is either dead or dying and will be unfit to eat.

It's best not to freeze clams in their shells as it tends to make them tough. To remove the shells, refrigerate the clams to make them easier to open. The cold relaxes the muscle that holds the shell tightly shut. Place a knife edge against the seam between the shells at the short, curved end of the clam. Slide the knife between the shells, without using the knife point. To release the clam from the shell, cut the muscle that holds the clam to the shells. Keep a bowl under the clams to catch the liquid. Wash the clams and drain them. Pack them with their own liquid and freeze them for up to six months.

When you want to serve the clams, thaw them in the refrigerator (a pint will thaw in six to eight hours) and cook them immediately. Cook just until the clams are heated through. It is easy to overcook them to a disappointing toughness.

To shuck a clam, hold the clam in your hand with the hinge toward you. Using a clam knife or dull table knife, slide the blade between the shell halves at the point opposite the hinge. Twist the knife to force the halves open, then slide the knife toward the back of the clam along the edge of one of the shells to cut the muscles. Twist off the top shell, then cut the muscles on the bottom half to completely remove the clam. You may wish to shuck the clams over a bowl to catch any liquid.

Soft-Shell Crabs and Hard-Shell Crabs

Crabs are an American specialty. Along the East coast, you can buy or harvest delicious live soft-shell crabs. These are blue crabs that have molted their hard shells and been harvested before their new shells could harden. They should be alive when you buy them.

To kill and clean a soft-shell crab, first stick a knife point between the crab's eyes. Lift its top shell by the tapered points and scrape out all evidence of the spongy intestines between the shell and the body. Remove the tail. Wash the crab thoroughly. Drain, wrap and freeze it for up to two months.

To prepare frozen soft-shell crabs, thaw them in the refrigerator (it will take six to eight hours), then immediately sauté them, broil them or fry them. Eat the crabs with your fingers, shell and all.

Hard-shell crabs, such as the Atlantic blue, West Coast Alaska king crab and Dungeness crab, involve very different freezing procedures from soft-shell crabs. You should freeze them cooked rather than raw, and once cooked, you should remove their shells, except for the claws, before freezing the crabmeat. The crabs you buy should be live. Cook them in boiling water that contains half a lemon,

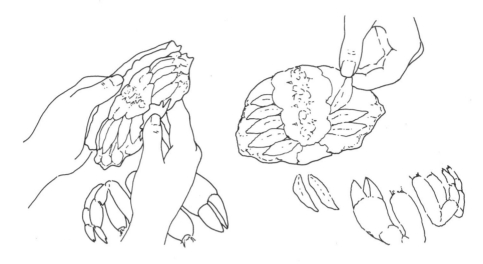

After removing the crab's legs, hold the crab with the cavity opening toward you and gently pry off the top shell with your thumb or a knife. Pull off the grayish white gills and discard. Scoop out the meat and pack.

some parsley, crushed peppercorns and a bay leaf. Boil them for eight minutes per pound. The crabs signal that they're cooked when they turn bright pink. Cool them quickly under cold water.

Snap off the two large claws. You can freeze the claw meat right in the shell, or crack open the shell with a nutcracker and remove the meat. Discard the legs unless they're large enough to contain some meat. Twist them inward and then pull them off. Pry off the top shell with a knife or your thumb. Remove the soft entrails under the shell. Discard the gills and any membranes. Pick out the crabmeat in the shell, and pack it snugly into freezer bags. Freeze it for up to three months. Thaw it in the refrigerator (it will take six to eight hours) before cooking.

Lobster

If you want to boil a lot of lobster meat with freezing in mind, it's most economical to buy large lobsters (6 to 8 pounds) rather than the smaller broiling lobsters.

Your lobster must be alive on the day you freeze it. If you're not sure that the lobster you're buying is alive, pick it up by the back and look at the tail; it should curl under the body. If it doesn't, the lobster is dead or dying. If it's dead, you have no way of knowing how long it's been that way, so don't buy it. Lobster must be fresh not only because it starts decomposing so rapidly but also because its wonderful sweetness fades away after one or two days of storage.

Cook lobster before you freeze it. You can plunge a live lobster directly into boiling water, but there is a more humane, less painful way to deal with it. To make it numb to pain, sever its spinal cord by cutting between the tail and the body. Then kill it by running the tip of a knife in the gap between the head and the shell of the front of the abdomen.

Then lower the lobster into boiling water. When it turns bright red, it's ready. Cool it in cold water. Split the lobster in half. Discard the stomach sac you'll find behind the head, and the black vein. Then pick out the meat and freeze it, or freeze the claws and tail intact. Lobster can be frozen for up to three months. You can thaw it in the refrigerator in six to eight hours.

You can also freeze whole lobsters if you have enough space to store them. To freeze a whole lobster, kill it and quickly plunge it into boiling water for a few moments. You can then simply cool it down, wrap it and freeze it, without cleaning. Do not thaw frozen whole lobster before cooking it.

Mussels

If you are gathering your own mussels, make sure it is in an area where the mussels are safe to eat. Mussels that you buy will be safe to eat. They are inexpensive and available year-round. Like clams, they should be shut tight if you're going to eat or freeze them. But unlike clams, mussels should be cooked before you freeze them.

Wash them very well, scrubbing off the crust on the shell and snipping off the stringy "beard." Steam them in a little water for a few minutes, just until they open. Strain their broth through several layers of wet cheesecloth to trap the sand. Pack them with their broth and with or without their shells. Mussels will freeze for up to one month, and thaw in the refrigerator in six to eight hours per pint.

Oysters

Buy large oysters if you intend to fry them after freezing. Buy small ones if you want to use them in stew after freezing. If you plan to eat the oysters within three days after you buy them, you can simply pack them in a container and hold them on ice in the refrigerator. Discard any whose shells are open.

Shellfish are delicacies, and they should be handled delicately. When they're overcooked, they become tough.

Treat oysters for freezing as you would clams. Freeze them raw, without their shells, and packed in their own liquid. To open oysters, which are far more resistant than clams, place the flat shell up (they have one shallow shell and one deep one) and insert a knife point between the shells near the narrow end where the shells meet. Cut through the muscle, and run the knife around the slit between the shells. Work with a bowl beneath the oysters to catch their juice. Opening oysters is hard work no matter how adept you are, unless you have a special, and rather expensive, oyster-opening device.

Rinse the shucked oysters and strain the liquid through several layers of wet cheesecloth to filter out the sand. (Set the pearls aside.) Package the oysters with their liquid and freeze for up to six months. Thaw them for six to eight hours per pint in the refrigerator.

Scallops

The scallops that we buy are actually the succulent muscle that the scallop uses to open and close its shell. Whether buying scallops out of the shell or removing them yourself, they should be packed in rigid containers, covered with water and frozen as soon as possible. Scallops freeze well for three to four months. Thaw them in the refrigerator for six to eight hours per pint.

Shrimp

Shrimp will be more tender if you freeze them raw. They get tough if you freeze them after you've cooked them. You can remove the shells before you freeze them or after you freeze them, before you cook them or after you cook them; it doesn't matter. In any case, wash the shrimp well to remove the sand and cut off the heads, if there are any. After you peel the shrimp remove the dark vein that runs down the middle of their backs, using a knife or your fingertip.

Freeze shrimp by laying them out on a cookie sheet and putting them in the freezer. When they are solid, collect them in a freezer bag. Use raw frozen shrimp straight from the freezer, but thaw cooked frozen shrimp in the refrigerator six to eight hours. Raw shrimp can be frozen for six months, cooked shrimp for four months.

CIOPPINO

Cioppino is a well-known seafood stew that did not originate in Italy, as many people suppose and its ingredients might indicate, but in San Francisco. It is a good way of using frozen fish and shellfish, but the fish should be barely cooked. For this reason, make the broth first and then add the fish, cooking it gently until just done. You might want to cook fish and shellfish in small, separate pots in broth from the soup and return them to the stew pot when they are done. You can use many kinds of fish in cioppino, but firm-fleshed fish such as monkfish work best because they retain shape and texture in the stew.

¾ cup olive oil
2 cups chopped onions
1 cup chopped green
    peppers
2 cloves garlic, minced
2 cups tomato puree
3 cups Fish Stock (page
    78), thawed if frozen
2 cups tomato juice
½ cup minced fresh
    parsley

1 teaspon dried thyme
2 bay leaves
    freshly ground black
    pepper, to taste
3 pounds mixed, frozen
    firm-fleshed fish, such
    as halibut, swordfish
    or monkfish
1 pound frozen shrimp,
    cooked with shells
    intact

Heat olive oil in a large pot over medium heat, and gently sauté onions, green peppers and garlic until the onions are soft. Do not allow the vegetables to brown.

Add tomato puree, stock, tomato juice, parsley, thyme, bay leaves and pepper and bring to a boil. Cook over medium heat for 30 minutes. Remove bay leaves, taste and adjust seasonings.

Cut the fish into serving-size pieces. Add to the stew along with the shrimp and cook over medium-low heat until the fish is just done. This could take from 15 to 30 minutes, depending on the sizes of the pieces of fish. It is best to check the fish frequently so it doesn't overcook.

Serve at once. Cioppino should be served with crusty bread and a salad.

YIELD: 6 to 8 servings

NOTE: Shrimp is usually left in the shell. If you prefer, you may substitute ½ pound cleaned shrimp for the shrimp in the shell.

CITRUS AND CELERY
PANFRIED FISH

Fresh fish can be frozen in the marinade, uncooked, for 6 months. Do not refreeze if made with frozen fish.

Marinade

¼ cup lemon juice
⅔ cup orange juice
¼ cup white vinegar
1 teaspoon dry mustard
2 teaspoons celery seeds
1 teaspoon grated orange peel

1 medium-size onion, chopped
8-12 (depending on size) small whiting or fresh sardines, thawed if frozen, pan-dressed

Breading

⅓ cup sesame seeds, ground
⅓ cup cornmeal

⅓ cup whole wheat flour
vegetable oil for frying

Sauce

reserved marinade
1 tablespoon honey

2 teaspoons cornstarch
2 tablespoons water

orange and lemon slices

celery tops

To marinate the fish, combine lemon and orange juices and vinegar in a bowl large enough to accommodate the fish. Dissolve mustard in about 1 tablespoon of the liquid, then stir back into the mixture in the bowl along with celery seeds, orange peel and onions. Add fish, turning several times so that all parts are coated by the marinade. The fish can be frozen now in the marinade, or cover and refrigerate for at least 8 hours, turning the fish occasionally.

Drain fish, reserving marinade. Combine sesame seeds, cornmeal and flour. Dip fish into breading, coating well on all sides. Fry in a shallow layer of medium-hot oil for 1 to 2 minutes on each side or until golden brown in color. Arrange on a serving dish and keep warm.

Pour the marinade into a small pot, letting in only some of the celery seeds for flavor and appearance. Add the honey and bring liquid to a boil. Dissolve cornstarch in water and slowly stir into the marinade.

Continue to cook, stirring until mixture is clear and thickened, about 2 minutes.

Serve sauce over fish, garnished with orange and lemon slices and celery tops.

To prepare frozen fish, thaw overnight in refrigerator, drain off marinade, then proceed with breading and cooking as above.

YIELD: 4 to 6 servings

SPANISH FISH SALAD

Make this refreshing salad with cod or any lean fish.

1¼ pounds frozen cod fillets,
    partially thawed
3 oranges
½ cup freshly squeezed
    lime juice
¼ cup peanut oil

2 tablespoons chopped
    fresh chervil or parsley
ground black pepper,
    to taste
Boston lettuce leaves

Place cod in a steamer and steam for 6 to 10 minutes, depending on the size of the fish. Fish should be opaque and cooked through.

Take zest from one orange, cut into short, thin strips and set aside. Peel the remaining two oranges. Remove the white skin and membranes from all the oranges, then cut through the membranes which separate the orange segments and put segments into a mixing bowl.

Mix together lime juice, oil, chervil and pepper. Break up the cod into the bowl with the orange segments, add the zest and pour the lime juice mixture over all. Toss carefully with your hands. Chill until ready to serve.

Arrange lettuce on a serving platter or in a shallow bowl. Mound cod salad in the center of the lettuce.

YIELD: 4 servings

SESAME BLUEFISH

An excellent way to enliven frozen fish fillets.

4 frozen bluefish fillets,
   about 1½ pounds total
   weight, thawed until
   just flexible
2 tablespoons sesame or
   olive oil

2 tablespoons lemon juice
2 tablespoons soy sauce
2 cloves garlic, minced
2 tablespoons sesame
   ·seeds

Place fish on broiler rack. In a small bowl, combine oil, lemon juice, soy sauce and garlic and spoon over fish. Sprinkle with sesame seeds. Broil the fish until opaque all the way through, about 15 minutes if fillets were partially thawed, 8 to 10 minutes if completely thawed.

YIELD: 4 servings

VARIATIONS: Substitute mackerel, trout or other fatty fish for the bluefish.

STEAMED FLOUNDER
WITH YOGURT-HORSERADISH SAUCE

Seasoning the steaming water, as in this recipe, is an excellent, low-calorie way to brighten the flavor of frozen fish.

1½ pounds frozen flounder
   fillets, partially thawed
1 cup chopped onions,
   thawed if frozen
2 stalks celery, chopped
¾ cup sliced carrots,
   partially thawed if
   frozen
1 clove garlic, minced

1 cup water
½ cup red wine vinegar
2 bay leaves
3 whole cloves
3 whole allspice
1 teaspoon cumin seeds
   Yogurt-Horseradish Sauce
   (page 61)

Arrange wet cheesecloth along bottom and sides of a steamer. On the cheesecloth, arrange fish, onions, celery, carrots and garlic. In bottom of the steaming utensil, combine water, vinegar, bay leaves, cloves, allspice and cumin seeds. Steam for 10 to 15 minutes.

Place fish and vegetables on heated serving platter. Serve with sauce.

YIELD: 4 servings

FLOUNDER WITH RED PEPPER SAUCE

Red, white and green, this dish is as beautiful as it is delicious. Serve it with something simple like brown rice or boiled potatoes, rather than something dramatic in flavor or bold in color that might compete with it on the plate.

¾ cup coarsely chopped
    red bell pepper
2 cloves garlic, chopped
1 tablespoon vegetable oil
    pinch of cayenne pepper
4 frozen flounder fillets,
    about 2 pounds

¼ cup half-and-half
2 tablespoons minced fresh
    parsley
1 lemon

In a large skillet, cook red pepper and garlic in oil until pepper is soft. Stir in cayenne. Puree the mixture in a blender or food processor and return to the skillet. Bring to a simmer and lay the fish fillets on top. Cover and cook over low heat about 10 minutes or until the fish is opaque all the way through.

    With a spatula, remove the fish to serving plates leaving as much of the pepper puree in the pan as possible. Stir the half-and-half into the puree and cook until slightly thickened and heated through. Spoon the sauce around, not on top of, the fish. Sprinkle the parsley on the fish (not on the sauce) and garnish each serving with a lemon wedge.

YIELD: 4 servings

CRAB SUPREME
IN GREEN PEPPER CUPS

Can be frozen for 3 months.

6 tablespoons butter
3 tablespoons Mochiko
   rice flour
2 cups half-and-half
2 tablespoons tomato
   paste
½ cup shredded Gruyère
   cheese
1½ teaspoons paprika
⅛ teaspoon cayenne pepper
2 tablespoons lemon juice
¼ pound mushrooms,
   finely chopped

¼ cup minced scallions
1 clove garlic, minced
¼ cup minced red bell
   pepper
12 ounces flaked crab meat,
   cleaned of cartilage
   and shell, thawed if
   frozen
5 medium-size green
   peppers

Melt 4 tablespoons of the butter in a 2-quart saucepan. Stir in rice flour with a wooden spoon until well blended. Cook over medium heat for 1 minute. Slowly add half-and-half, whisking until the mixture thickens slightly. If it thickens too much, thin with more half-and-half. Whisk in tomato paste, cheese, paprika, cayenne and lemon juice. Remove from heat.

Melt the remaining 2 tablespoons butter in a small skillet. Add mushrooms, scallions, garlic and red pepper. Sauté gently until any juices given up by the vegetables have evaporated but do not let the vegetables brown.

Stir sautéed vegetables and crab meat into sauce. Taste and adjust seasoning. The sauce can be frozen at this point for later use.

If making this dish to serve immediately, cut green peppers in half lengthwise and remove core and seeds. Bring 2 quarts of water to a boil and blanch pepper halves until just tender, about 4 minutes. Remove, drain and transfer to heated plates. Gently heat sauce and ladle into green pepper cups. Serve with brown rice.

To use frozen sauce, thaw in the refrigerator overnight, then reheat, stirring constantly, while the green pepper cups are blanching.

YIELD: 5 servings

NOTE: The blanched pepper cups can be filled with the crab meat sauce and frozen. When ready to prepare, bake loosely covered in a 375°F oven for about 35 minutes, or until hot and bubbly.

VARIATIONS: Instead of halved green peppers, use tiny, immature peppers (available in the fall) or avocado halves.

RED SNAPPER FILLETS STEAMED WITH FENNEL

Freeze the feathery fennel greens needed for this recipe while they're in season. This preparation would also be good with bluefish. The steaming is done on a platter set on supports in a covered wok or tiered Chinese steamer. A large skillet, with a platter set on a cooking rack, would also work.

4 frozen snapper fillets, about 2 pounds, partially thawed
¼ cup finely chopped fennel leaves, or 1 tablespoon dried fennel

2 cups chopped, seeded tomatoes, thawed if frozen
3 tablespoons butter
2 tablespoons rice wine vinegar

Lightly butter a rimmed, ovenproof dish large enough to hold fish in a single layer but small enough to fit into a wok or steamer.

Place fish on dish. Cover with fennel and tomatoes. Dot with butter and sprinkle with vinegar.

Place dish on steamer rack, and steam above boiling water for about 15 minutes, until fish flakes easily.

Carefully remove plate from steamer rack to avoid spilling accumulated liquid.

Pour off liquid into a small saucepan. Boil over high heat to reduce liquid to ¼ cup. Pour liquid over fish and serve.

YIELD: 4 servings

SCALLOPS PROVENÇAL

1½ pounds frozen sea
  scallops, thawed
½ cup whole wheat flour
2 tablespoons olive oil
4 teaspoons minced
  shallots
2 cloves garlic, minced
2 tablespoons finely
  chopped fresh parsley

1 teaspoon chopped fresh
  tarragon, or ½ tea-
  spoon dried tarragon,
  crushed
1 lemon, halved
  hot pepper sauce,
  optional

Near serving time, slice the scallops in half widthwise, dredge in flour and broil for 2 to 3 minutes.

Heat frying pan to medium and add oil. Add scallops, shallots, garlic, parsley and tarragon. Sauté for 2 minutes. Squeeze lemon over scallops. Season with a dash or two of hot pepper sauce, if desired.

YIELD: 4 servings

VARIATION: Cut any cooked, firm-fleshed fish into bite-size pieces and sauté with the herbs and olive oil, as described above, just until hot.

CURRIED SHRIMP
AND LEEKS IN PAPAYA

Hot brown rice is the best accompaniment for this dish. The prepared curry should not be frozen.

½–¾ pound frozen peeled
     cooked shrimp, thawed
2 cloves garlic, minced
2 tablespoons peanut oil
2 cleaned fresh leeks or
    2 frozen leeks, partially
    thawed
2 tablespoons butter

2 teaspoons curry powder
⅓ cup milk
2 papayas, halved and
    seeded
    pineapple or avocado
    slices
1 lime

Cook shrimp and garlic in oil just until the garlic is fragrant, about 2 minutes. Set aside.

Quarter the leeks lengthwise. Cook leeks in butter until tender, about 8 minutes for fresh leeks, 4 minutes for frozen leeks. Stir in the curry powder, adding a spoonful or two of water if necessary to keep it from sticking. Add milk and shrimp and cook about 30 seconds. Divide hot shrimp mixture among papaya halves, garnish with pineapple or avocado slices and serve each with a lime wedge.

YIELD: 4 servings

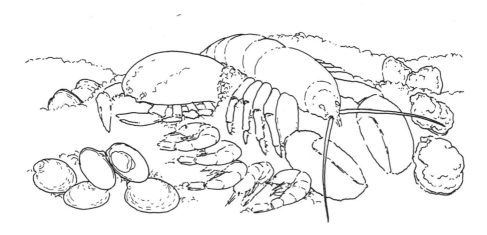

CHAPTER 8

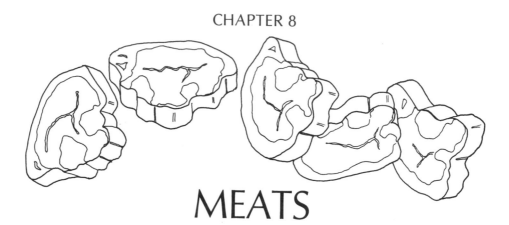

MEATS

When meat prices go up so do sales of new home freezers, and for good reason. Owning a freezer allows you to use the erratic pattern of meat prices to your advantage, stockpiling meat when it is inexpensive to consume when it is dear. There are many factors that cause meat prices to fluctuate—industry strikes, droughts in the Midwest, federal programs—and all of them are beyond the control of consumers. Except, that is, consumers with freezers who buy meat in bulk.

Buying in Bulk

Buying large quantities of meat for your freezer is not something to do on an impulse. Take the time to learn about meat, and think through your family's meat preferences and how much of any particular kind of meat you are likely to consume. Then do some comparison shopping. Not only will bad decisions cost you money, you'll also have to eat all those cuts of meat, even if your family grows tired of them.

There are many outlets to choose from for buying meat in bulk. You can call around to butchers until you find one who offers 10 to 20 percent discounts on orders over 50 dollars. Many butchers will sell you, for instance, a whole loin of pork instead of individual chops, then cut the meat for you and wrap the pieces individually. Some butchers actually specialize in selling bulk meats that they buy from farmers or large meat packagers.

There are also wholesalers and meat distributors who do nothing but sell whole sections of an animal, a quarter of a steer, for instance. In rural areas, you can go directly to a slaughterhouse and processing plant, where cattle and pigs are butchered, processed and packaged right on the premises. You can also bring your own animal to them for these services. Some will charge a "kill fee" and some will not. All will charge by the pound to cut and wrap the meat.

A recent innovation in bulk meat buying is taking place at the supermarket. Markets are now selling vacuum-packed, fully trimmed, boneless sections of meat. These packages, which weigh between 5 and 30 pounds, are called sub-primals because they contain a smaller section of a large wholesale cut, known as a primal. A beef carcass, for instance, is usually divided into four or five major cuts, or primals. The sub-primals typically available are the whole top round, the whole bottom round and the whole round tip.

Well-marbled meat is juicy and tender because the fat melts into the meat as it cooks. An abundance of marbling elevates the price of the cut, but it also elevates the amount of saturated fat in your diet.

The best bargain is the whole bottom round because it is the section that yields the eye of the round (a very expensive cut) plus a rump roast and ground beef of excellent quality. The whole bottom round will cost you as much per pound as you would ordinarily pay for ground beef. Sometimes the supermarket will cut and rewrap a sub-primal for you for free or for a minimal charge of 10 cents a pound. If the market won't do the cutting for you and you don't know how to do it yourself, get a manual for meat cutting. In either case, sub-primal cuts are a bargain, because you are buying them for significantly less than you would pay for the same meat divided up and sold at retail in the same store.

Sub-primals do, however, take a bit of getting used to. Because the meat is vacuum packed as soon as it is butchered, it has a sort of purple color until you open it and let the air get to it. Exposure to air will change the meat very shortly to its characteristic red color. (Critics believe that beef packaged this way is insufficiently aged and is therefore not as tender or flavorful. Beef sold at retail is aged from six to ten days simply in the course of bringing it from the packer to consumer. Controlled aging is usually reserved for expensive rib and loin cuts of beef and lamb that are sold to restaurants. Veal and pork are very tender and are never aged.) Vacuum-packed meat also has a sour odor when you first open it, but it dissipates within an hour. If it doesn't, of course, you should return the meat.

You can also buy in bulk merely by buying quantities of your favorite cuts when they go on sale at the supermarket. This is the simplest but least economical way to stock up on meat unless you buy large pieces that you divide at home into smaller parts. It does have the advantage, however, of letting you see exactly what you are buying.

Know What You're Buying

The U.S. Department of Agriculture employs a very useful system of grading meat for quality and yield (the percentage of good retail cuts a carcass can be divided into). While the system is a very reliable guide, it is a voluntary one, and only about 50 percent of the meat sold is graded. The grading is done at the request of the slaughterhouse. Grades are awarded on the basis of the appearance of the carcass and are stamped on it in harmless purple vegetable dye. For beef, the system lists several quality grades, because beef quality varies more than any other meat.

The highest grade is Prime, and it is awarded to only 5 percent of beef. Though some Prime beef, lamb and veal is sold at retail, most of it goes to restaurants.

Choice is second to Prime in quality, has slightly less marbling of fat but is also juicy and tender. Good and Standard grades, which come next on the scale, have considerably less marbling than Prime or Choice. Marbling is not necessarily a good thing, since those streaks and flecks of fat are full of cholesterol. Good and Standard grade meats require slow cooking in liquid, but they're full of flavor and less expensive than the higher grades.

The yield system, graded 1 through 5, is based on the ratio of good lean meat to waste and fat. Yield 1 meat can realize about 80 percent retail cuts; in other words, 80 percent of the meat is usable. Yield 2 meat produces 75 to 80 percent retail cuts, Yield 3 produces 70 to 75 percent retail cuts, Yield 4 produces 66 to 70 percent, Yield 5 less than 66 percent.

Only 5 percent of the beef sold commercially meets Yield 1 standards. Twenty-five percent is graded Yield 2 and 50 percent is Yield 3. Find out the yield, if you can, when you buy meat for the freezer. A carcass with a higher yield justifies paying a higher price for it. If there's no difference in price, then always buy the meat with the higher yield. The Federal Trade Commission advises consumers to always ask to see the shield-shaped USDA stamps for quality and yield.

Whether you buy a quarter, a side or a whole carcass, be sure you know how much meat you are actually getting. Bulk meat is sold at the hanging weight, which is the weight of the entire uncut carcass. Obviously, once the meat has been cut and trimmed it is going to weigh considerably less, sometimes less than half of its bulk weight. Make sure that you know the cost of a pound of edible meat and don't confuse it with the cost of a pound from the uncut carcass. Ask specifically how much the meat costs when wrapped and frozen.

Find out exactly what cuts you are going to wind up with, too. Pick only what you can use and only what you have room for. For every cubic foot of freezer space you have, you will be able to store roughly 35 to 40 pounds of wrapped meat. If you buy a quarter of a steer, a great deal of it will come in the form of bones and stew meat that may exceed your need for such parts. Here's what to expect: If you were to buy a whole beef carcass, which is unlikely because it would weigh about 420 pounds, you could expect that 40 percent of it would be steaks and roasts, 20 percent would be pot roasts and 20 percent would be stew meat and ground meat. The rest would be surplus fat and bone.

If you bought the forequarter of beef, the trimmed yield would be about 140 pounds. The forequarter contains the greatest ratio of waste (fat and bone) to meat. Twenty-five percent of the trimmed yield would be in the form of steaks and oven roasts, and 69 percent would be pot roasts, stew meat and ground meat.

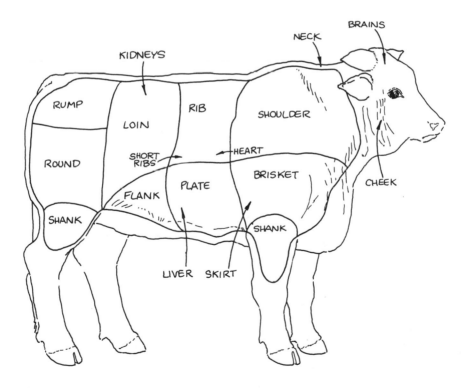

Beef cuts.

The hindquarter is a choicer cut of meat. From 140 pounds of trimmed hindquarter, you would get 58 percent steaks and oven roasts and 18 percent stew meat, ground meat and pot roasts.

When you buy bulk meat, it's important to understand animal bone structure. Beef, pork, lamb and veal have the same bone structure and the same names for those bones. The parts of the animals that move a lot and get a lot of exercise are the least tender—the shoulders, the legs, the flanks and the breast. The muscles that are least often used are the most tender, but they comprise a distinct minority of the animal's flesh. The tender parts of the animal are along the backbone in the rib and loin section.

Ask Questions Before You Buy

Comparison shop if you want to buy meat in bulk, then if you have a choice of several reputable places, choose the cheapest and the most cooperative. A cooperative meat supplier will have someone knowledgeable around for you to talk to and will honor your special requests. Ask a lot of questions.

If you want some of your beef or pork bulk purchase to be in the form of processed meat, that is, meat that will be cured or dried, cooked or seasoned, ask how it will be done. There are hundreds of varieties of processed meats and sausages and you will have to inquire which, if any, your supplier is prepared to produce. We do not advocate the consumption of cured meat, but if it is to be part of your order, ask how it will be cured. Salt and honey or brown sugar are far better than a salt and nitrate grind. In addition, make sure that smoked meats are really smoked and not just flavored with liquid smoke.

Ask the supplier how the meat is frozen. Meat that is flash-frozen at about −30°F will have the smallest ice crystals, and its texture, color and flavor stays closest to fresh. Check that the butcher will package the meat in quantities convenient for you to use, will use good quality freezer paper and will wrap the packages with care. Ask to have patties, chops and steaks separated with double layers of freezer paper. And ask to have all the packages labeled carefully. Ask the butcher to freeze the meat for you. If you undertake to freeze a large quantity of meat at one time, you will overwork your freezer and perhaps even cause it to fail. Beef in the form of roasts or other large pieces can be frozen up to twelve months. Ground or cubed beef freezes for three months.

Lamb

Lamb does not vary in quality to the same degree that beef does. Supermarket lamb is usually Choice, while butcher shops get Prime lamb. Once a lamb has its first birthday, it becomes mutton, of which very little is available. Since lambs are

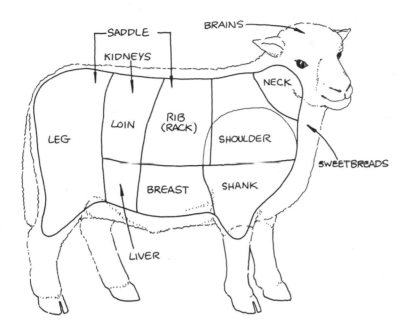

Lamb cuts.

so small, you can freeze a whole carcass, and its yield of usable meat will only run from 35 to 40 pounds. Of that, about 20 to 25 percent will be lost in trimming it, but the savings of buying a whole lamb rather than buying the same cuts retail is still 20 to 25 percent.

Once a lamb has been butchered for freezing, there will be no leftover odds and ends like bones or small pieces that you can grind. If the total yield is, say, 38 pounds, 31 pounds will be legs, chops and shoulders, and 7 pounds will be breast and stew meat. All the bones will still be in the cuts. Lamb also has brains, heart, kidneys, liver, sweetbreads and tongue. Two kidneys will come with your order, but the other parts won't unless you order them in advance.

If you know generally how you will use the lamb, you can have it cut into basic pieces that you will be sure to use. You should discuss the intricacies of each part of the lamb and the many possibilities for cutting it with your supplier.

Greek, Italian and the best American butchers can supply you with a whole baby lamb in the spring, if you order it in advance. For the most part, this lamb comes from the grazing lands of Colorado and Utah. If it is slaughtered between March and October it is called genuine spring lamb. The fact that lamb has been allowed to graze is your assurance that it tastes good because it has fed well.

If you buy lamb in the supermarket, check the expiration date on the package. Don't buy packages that contain liquid. The liquid is an indication either that the lamb has been in the case for some time or that it was frozen and thawed.

Ground lamb can be made from any cut of the carcass. The ground lamb in the supermarket, therefore, could have come from any part of the animal, and more than likely it will have come from one of the fattier parts. It's more economical to buy a shoulder of lamb, trim the fat and grind it yourself or have the butcher do it for you.

Lamb roasts or other large pieces can be frozen for six to nine months. Ground or cubed lamb freezes for three months.

Pork

Pork does not freeze as successfully as beef or lamb. It tends to harden in the freezer, because most of its moisture is in its fat rather than in its flesh. To maximize its freezer storage life, you should first cut away as much of the fat and bone as you can. Even with careful trimming, you ought to freeze pork roasts or other large pieces no longer than eight months. Pork Chops will freeze for three to four months, and ground or cubed pork for one to three months.

The pork you buy is certain to be tender because it comes from pigs slaughtered when they are young. The uniformly high quality of pork makes a grading system unnecessary. There is a 1, 2, 3 numbering code for pork, but it is not used much.

While freezing is, in general, no great boon to pork fanciers, it does perform one very useful service. It kills the parasites which can hide in pork flesh and

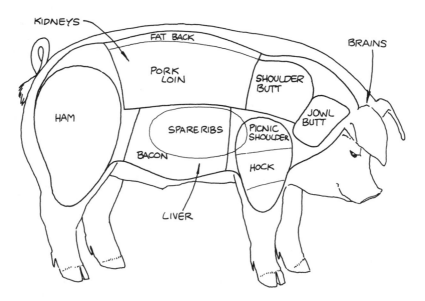

Pork cuts.

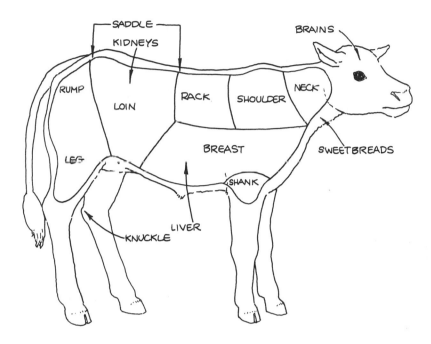

Veal cuts.

cause the disease called trichinosis. The parasites are destroyed by freezing pork at 5°F for 20 to 30 days or at −10°F for 10 to 20 days or at −20°F for 6 to 10 days. The only other way to kill the parasites is to cook the meat to an internal temperature of 170°F. The temperature often suggested for cooking pork, 185°F, is really higher than necessary and dries out the meat.

Veal

Beef and veal both come from cattle. Veal is between one and three months old. When it's between 7 and 10 months old, it's a calf and it's known as baby beef. Once matured, between 15 and 30 months, it's ready to be sold as beef. Even though baby beef comes from an older calf than veal, it too is very perishable.

Veal, like pork, is very tender. It has almost no fat and little firmness of texture. Because it is the most expensive meat, many people do not freeze veal, but prefer to buy it fresh and use it immediately. Buying veal or calves in carcass form is not a good idea because they lose their moisture, and along with it their flavor and texture, in the freezer.

If you do want to keep veal in the freezer, freeze only large cuts. If you freeze the very thin veal slices (or scallops) that are used for veal scalloppini, their thinness would make them highly susceptible to freezer burn. Instead, buy a large piece of the leg and cut scallops or steaks from it when you thaw it. You can freeze veal steaks, however, if they are at least ¾ inch thick. You can also freeze veal that you intend to grind, or grind it and then freeze it. Freeze ground veal and veal for scalloppini for one month at the very longest. Freeze large pieces of veal or veal chunks for stew that were cut from the breast, neck or shank for no more than three months.

Game

All game meat, wild and domestic, can be frozen. Wild game should at least cool naturally before you freeze it. If you leave your game in the freezer for six months, the freezer will do the job of tenderizing it for you. If you intend to eat your game sooner than that, and it's any bigger than a rabbit, you ought to hang it first to tenderize it before you freeze it. The freezer needs a long time to get the job done. Freeze hearts and brains and livers, even the scraps and suet. The scraps can serve to feed birds and pets. Add in the suet when you grind lean meats. Cooked game can be frozen for two months.

Rabbit is becoming increasingly available cut up and frozen in supermarkets or fresh from butchers. You can cook it in all the ways you cook chicken, and its flavor is more interesting than that of chicken.

Variety Meats

Your diet may well lack the diversity of variety meats. When you buy a quarter or a side of an animal, ask for the liver, heart, kidneys, brains and sweetbreads. In the refrigerator they'll last two days at most, but in the freezer they'll last up to a month, except for liver, which has a storage life of four months.

Cut liver for the freezer into slices at least ½ inch thick for frying or broiling. Package it with double layers of freezer paper between the slices.

Trim the heart of any hard areas of muscle and large arteries. Rinse, drain and wrap it. Use it in stuffings and sauces. It is tough, being a well-used muscle, and will require long, slow cooking.

Trim the fat from kidneys and freeze them whole. Enjoying their strong flavor is an acquired taste, and for the most part Americans have preferred not to acquire it. If you soak kidneys in water with a little lemon juice for half an hour before freezing them, you can calm down their pungence a bit.

Trim the excess blood clots and membranes from brains before freezing them. Brains must be very fresh when you buy them and quickly frozen when you get them home. Soak them for one hour in 1 quart of cold water that contains a little lemon juice. Package sliced, diced or whole. Sauté them in butter to serve.

Sweetbreads are the two lobes of the thymus gland of very young animals. They are as perishable as brains, so much so that they actually disappear in older animals. Soak them in cold water with a little lemon juice for two hours. Trim the extra membrane and freeze. You can also freeze tongue and tripe, though they are usually pickled.

Marrow

All bones contain marrow. The marrow you use for cooking comes from the long leg bones. The marrow is much easier to remove if you ask the butcher to saw beef shank bones into segments that are about 4 inches long. Use a small, sharp knife to scratch the marrow out of the bones. Wrap it tightly in plastic wrap and then overwrap it by putting it in a freezer bag. It will freeze for two months.

Fresh Sausage

Fresh sausage is not smoked or cured. It is made from uncooked meat and you should therefore treat it as you would any raw meat. Fresh sausages can be made entirely of pork or beef or a mixture of meats. If you make your own sausages, you can keep them in the refrigerator for two days. If that's not time enough to eat them all, wrap each individually, pack them into a freezer bag and freeze for two to three months.

Wrapping Meat for the Freezer

Meat will eventually dehydrate in the freezer in spite of the most careful wrapping. But there are some measures you can take to retain maximum quality in your frozen meat. Aluminum foil is likely to be torn by odd shapes and projecting bones, so wrap your meat in freezer paper. Make the wrappings as airtight as you can by pressing the wrap tightly around the meat and sucking out the air with a straw. Wrap meat chunks in a single layer in recipe-size portions so they will freeze evenly.

To wrap patties of sausage or ground meat, cut a long strip of freezer paper twice as wide as the patties. Fold the paper in half lengthwise with the dull sides

Ground meat patties store best stacked for easy removal. Start with a long piece of freezer wrap, folded in half with the shiny side facing out. Lay one patty on the end of the wrap and fold the wrap over the top. Continue layering meat patties and wrap, then place the stack in a plastic freezer bag and seal.

facing in. Place one patty on one end of the paper and cover the patty by folding the paper strip up over it to cover it. Center a second patty on top of the first and cover it. Then put all the patties in a freezer bag, press out as much air as possible, seal and freeze. Steaks and chops, too, are safer from freezer burn if you wrap several at a time, separating them with double layers of freezer paper.

When freezing oddly shaped pieces of meat, crumble up some freezer paper and use it to pad projecting bones. Wrap the meat tightly. Label all meat with the cut, the weight and the date.

Storing Meat in the Freezer

While it is certainly a waste of time to fill up a freezer with loaves of bread, it is a waste of money to fill the freezer with more meat then you can use within the optimum storage time. Or to stock so much meat that you find yourself eating it much more frequently than you ordinarily would. It takes a very long storage time for meat to become unsafe to eat because the bacteria cannot multiply at 0°F. You will, however, be eating meat of considerably diminished quality if you keep it frozen for longer than suggested. It won't taste as good, and it won't be as juicy or as nutritious. The enzymes, which are slowed but not stopped by freezing, and the air inside the wrappings will turn the meat rancid.

The rate of rancidity varies from animal to animal. Since beef and lamb contain a lot of saturated fat, their storage life is considerably longer than that of pork. Pork fat contains unsaturated fats, which when mixed with oxygen turn rancid. You can definitely tell if meat has become unsafe to eat. If you open a package of meat to discover that it has a gray color or an off odor and a slippery surface, throw it out.

Thawing Meat

Thaw meat in the refrigerator still sealed in its original wrappings to prevent moisture loss and possible contamination. If you are tempted to thaw a thick cut of meat on the counter because you've forgotten to take it out of the freezer soon enough, think again. The surface of the meat will thaw long before the inside does, allowing plenty of time for bacteria to grow. If you're really in a hurry, and your meat is thick, slip the meat into a waterproof bag, seal it tightly and leave it in a sink full of cold water. The same procedures apply to thawing ground meat. Here is a summary of thawing times in the refrigerator for various kinds of meat. Large pieces of beef will thaw in 4 to 7 hours per pound; roasts and smaller pieces in 3 to 5 hours per pound. A 1-inch-thick steak will thaw in 12 to 14 hours. Ground or cubed beef takes 10 to 12 hours per pound to thaw.

Large cuts of lamb need 4 to 7 hours per pound to thaw; roasts and smaller pieces take 3 to 5 hours per pound. Ground or cubed lamb thaws at a rate of 10 to 12 hours per pound.

Large pieces of pork thaw in 4 to 7 hours per pound; roasts and smaller pieces in 3 to 5 hours per pound. Pork chops thaw in 24 hours, ground or cubed pork in 10 to 12 hours per pound, and fresh sausage in 8 to 10 hours.

Veal roasts and pieces need 3 to 5 hours per pound to thaw; ground or cubed veal takes 10 to 12 hours per pound.

Marrow thaws in 7 to 8 hours, and tongue in 12 to 14 hours.

Wild game, cooked or uncooked, takes 48 hours to thaw.

Variety meats should be thawed in the refrigerator for 12 to 14 hours, on a plate to catch the drips. Or you can cook them directly from the freezer, allowing one-third to one-half longer cooking time.

Once meat has defrosted, you can keep it in the refrigerator as long as you would fresh meat, one to two days for ground meat and variety meats, two days for stew meats, three days for ribs, steaks and chops and three to five days for roasts. This is a matter of some controversy; some people say thawed meat should be in the refrigerator two days at most. In any case, since the meat has lost moisture, the sooner you cook it, the better.

You can cook meat without thawing it, but naturally the cooking time will be longer. Steaks and chops will take 1½ to 2 times as long to cook frozen as it would take to cook if you thawed it first. Roasts will take roughly 1⅓ to 1½ times as long as when thawed. In addition to the time factor, there are other reasons why it's better to thaw meat before cooking.

Because of frozen meat's solid nature, it is difficult to season it well, as any herbs or spices you add will not penetrate the meat, but simply fall off into the pot. It's also impossible to insert a meat thermometer into frozen meat to help determine its doneness.

KOREAN BEEF

Can be frozen for 2 weeks in the marinade or for 1 month if cooked.

1 pound flank steak,
  partially thawed if
  frozen
2 scallions, including green
  tops, thinly sliced
2 cloves garlic, minced
2 tablespoons sesame
  seeds, toasted and
  crushed (page 54)

3 tablespoons sesame oil
2 tablespoons soy sauce
½ teaspoon freshly grated
  gingerroot, optional

Cut steak across the grain into ½-inch slices. Cut the slices into 3-inch-long strips.

In a medium-size bowl, combine the scallions, garlic, sesame seeds, oil, soy sauce and gingerroot.

Add steak strips and toss. At this point, the beef and marinade can be frozen for cooking later. Or cover and marinate in the refrigerator overnight or for 2 hours at room temperature, turning steak pieces occasionally.

Thread meat onto skewers. If meat has been frozen in marinade, be sure it is completely thawed before cooking. Place on broiler pan. Broil on the highest oven rack until browned, 2 to 3 minutes on each side.

Serve the meat on the skewers or remove it. The cooked meat can be frozen at this point and reheated later. Best served with hot brown rice.

To heat frozen beef, thaw in the refrigerator overnight, then place in a baking pan and cover with foil. Bake in a 350°F oven for 10 to 15 minutes, checking every 5 minutes to make sure meat is not drying out. Add a few tablespoons of water to the pan if the meat seems dry.

YIELD: 4 servings

GREEN VEGETABLES WITH BEEF

Serve this as a main dish with brown rice or Oriental noodles such as rice sticks. Fresh or frozen vegetables and beef can be used in this dish. It can be refrozen for up to 6 months for later use.

2 tablespoons plus 2 teaspoons soy sauce
3 teaspoons Mochiko rice flour
2 slices fresh gingerroot, minced
1 clove garlic, minced
½ pound beef tenderloin or sirloin, partially thawed if frozen, cut into bite-size cubes
4 tablespoons cold water or chicken broth

2 tablespoons vegetable oil
2 cups bite-size pieces (any combination) green beans, broccoli, asparagus, snowpeas or green bell peppers, partially thawed if frozen
2 tablespoons minced fresh scallions

Make marinade by combining 2 tablespoons soy sauce, 1 teaspoon rice flour, gingerroot and garlic. Mix with meat and let stand 10 minutes before cooking.

Meanwhile, prepare a gravy in a separate small bowl by mixing remaining rice flour with 1 tablespoon of the water or broth. Stir until smooth. Then add remaining water or broth and 2 teaspoons soy sauce. Mix thoroughly.

Heat wok. Add oil and heat. Add drained meat and stir-fry quickly, about 3 minutes. Add gravy, stir-frying until thickened, about 1 minute. Add vegetables and stir-fry until vegetables are crisp-tender, about 3 minutes.

Sprinkle with scallions and stir. Adjust seasoning and serve immediately, or cool and freeze.

To make the frozen dish, thaw overnight in the refrigerator, then reheat slowly on top of the stove.

YIELD: 4 servings

VARIATION: Substitute ½ pound flank steak, partially thawed if frozen, for the tenderloin. Cut the steak into thin strips, and marinate 15 minutes. Proceed with recipe as directed above.

BEEF POTPIE

Can be frozen for 6 months.

½ cup plus 2 tablespoons
   whole wheat flour
¼ teaspoon ground black
   pepper
2 pounds lean chuck,
   partially thawed if
   frozen, cut into 1-inch
   cubes
¼ cup vegetable oil
½ cup chopped onions
2 cloves garlic, minced
4 cups water
1 bay leaf
3 medium-size potatoes
4 medium-size carrots

¼ cup chopped fresh
   parsley
2 teaspoons fresh thyme
   leaves, or 1 teaspoon
   dried thyme
2 teaspoons fresh basil, or
   1 teaspoon dried basil
2 tablespoons Worcester-
   shire sauce
1 tablespoon soy sauce
1 recipe whole wheat
   pastry (page 329)
1 egg
1 tablespoon milk

In a small bowl, combine ½ cup of the flour and pepper. Add the beef cubes and toss to coat. Shake off excess flour.

In a 6-quart pot, heat the oil over medium heat. Add 8 to 10 beef cubes and brown on all sides. Place browned cubes in a medium-size bowl. Repeat with remaining beef. Add the onions and garlic to the remaining oil in the pan. Sauté until soft, about 4 minutes.

Add water, beef, any liquid at the bottom of the bowl and bay leaf. Bring to a boil.

Reduce heat to low. Simmer, partially covered, for 30 minutes.

Meanwhile, cut potatoes into 1-inch cubes. Cut carrots crosswise into ½-inch slices. Stir potatoes, carrots, parsley, thyme, basil, Worcestershire sauce and soy sauce into beef mixture.

Simmer, partially covered, until vegetables are tender, 20 to 30 minutes. Remove about ¼ cup liquid and whisk in 2 tablespoons of the flour. Return to pot and simmer until thickened. Remove from heat and pour into a 3-quart ovenproof casserole dish. At this point, the beef mixture may be cooled and then frozen for later use.

Roll out pastry dough 1 inch wider than diameter of casserole dishes. Cut and secure dough to rim according to directions for Maine Blueberry Pies (page 268). Cut three 2-inch-long slashes across the crust.

In a small bowl, lightly beat together the egg and milk. Brush mixture onto crust. Bake at 375°F until crust is golden brown, about 45 minutes. Serve directly from casserole dish.

To prepare frozen casserole, cover frozen filling with top crust and proceed as directed. However, when crust is golden brown, cover pie with foil. Turn oven down to 350°F and bake until filling steams, about 45 minutes more.

YIELD: 6 to 8 servings

ORANGE BEEF

This recipe, utilizing round steak and orange juice concentrate, requires at least 4 hours to marinate the meat. But fast cooking in a wok (10 to 15 minutes) makes this main dish easily adaptable to busy schedules. The dish can be frozen for 4 months.

1 pound round steak, partially thawed if frozen
½ cup frozen orange juice concentrate, thawed
¼ cup honey
2 tablespoons soy sauce
4 teaspoons minced fresh gingerroot
peel of 1 large orange, divided into penny-size segments
⅓ cup vegetable or peanut oil

Slice steak across the grain into ½-inch strips.

In medium-size bowl, stir together orange juice concentrate, honey, soy sauce, gingerroot and orange peel. Add steak. Cover and marinate in the refrigerator from 4 hours to overnight, stirring occasionally to keep meat coated.

Heat oil in a wok. Remove beef from marinade with a slotted spoon and stir-fry until browned, about 5 minutes. Add remaining marinade.

Reduce heat to simmer and cook until marinade thickens, 5 to 10 minutes. Serve immediately over brown rice, or cool and freeze.

To make the frozen dish, thaw in refrigerator overnight and heat gently on top of the stove.

YIELD: 4 servings

FLANK STEAK STUFFED
WITH SPINACH AND MUSHROOMS

This dish can be served hot or cold. The same stuffing can also be used for poultry if you add ½ teaspoon of sage to the bread crumbs. The stuffed steak can be frozen for 3 months after cooking.

1½ pounds flank steak, partially thawed if frozen	1 tablespoon chopped fresh parsley
2 tablespoons butter	1½ cups bread crumbs
1 small onion, chopped	¼ cup chopped blanched spinach, thawed if frozen, well drained
1 small stalk celery, chopped	1 tablespoon vegetable oil
½ cup sliced mushrooms	½ cup water

Tenderize meat by pounding the steak on both sides with a meat mallet.

Melt butter in a small saucepan. Add onions, celery and mushrooms and sauté until tender, about 6 minutes.

In a medium-size bowl, toss together onions, celery, mushrooms, parsley, bread crumbs and spinach. Spread the stuffing mixture on the steak. Roll lengthwise as for a jellyroll. Tie with string at 2-inch intervals.

Heat oil in a large skillet and brown the stuffed steak on all sides.

Pour water into a roasting pan. Place the browned steak in the roaster, cover and bake at 350°F until tender, about 1½ hours.

Freeze now or remove the string, cut steak into 1-inch slices and serve with drippings. To serve cold, slice thinly and serve with Mustard Sauce (page 53).

To use frozen steak, thaw completely in the refrigerator overnight. Then bake with a little water in foil in a 350°F oven for about 20 minutes or until hot.

YIELD: 4 to 6 servings

MARINATED BEEF AND
MUSHROOMS

Uncooked beef can be frozen in this marinade for 2 weeks. The cooked dish can be frozen for 3 months.

1½ pounds round steak,
        partially thawed if
        frozen
2 tablespoons minced
        onions
2 cloves garlic, minced
2 tablespoons chopped
        fresh parsley

1 tablespoon lemon juice
2 tablespoons vegetable oil
2 teaspoons Worcestershire
        sauce
2 teaspoons soy sauce
¼ cup water
3 tablespoons butter
1 cup sliced mushrooms

Cut steak across the grain into ½-inch strips. Place in a shallow bowl.

In a small bowl, combine the onions, garlic, parsley, lemon juice, oil, Worcestershire sauce and soy sauce. Pour over beef and toss. Cover and refrigerate at least 2 hours or freeze in marinade for later use.

Place water and marinated steak in baking pan. Cover with foil and place in a 325°F oven. Bake until tender, 45 to 60 minutes. Remove beef from pan to a serving plate and keep warm. Pour drippings into a large skillet.

Add butter to drippings and heat over low heat. Add mushrooms and sauté until tender, about 5 minutes. Pour sauce and mushrooms over steak and serve, or freeze for later use.

To prepare the frozen dish, thaw overnight in the refrigerator, then heat slowly, covered, on top of the stove.

YIELD: 6 servings

SAUERBRATEN WITH GINGER GRAVY

Serve this dish with hot noodles, red cabbage, potato dumplings and ginger gravy. The beef should not be frozen for more than 1 week in the marinade.

¼ cup vegetable oil	1 tablespoon ground ginger
1 cup chopped onions	5 pounds bottom round
2 tablespoons mixed	beef roast, thawed if
pickling spices	frozen
1 cup red wine vinegar	½ cup water
2 cups apple juice	2 tablespoons whole wheat
¼ cup honey	flour

Place a doubled, large freezer bag in a large bowl. Add oil, onions, spices, vinegar, apple juice, honey and 1 teaspoon of the ginger. Knead and shake bag gently to mix marinade. Carefully add beef and mix again. Seal bag, label and freeze for 1 week.

Thaw in refrigerator for 2 to 3 days. Remove meat from marinade and place in a roasting pan. Strain marinade. Pour strained marinade over beef and roast, covered, at 350°F for about 2 hours, until roast has an internal temperature of 165°F. Baste occasionally with marinade. When ready to serve, remove the roast from the pan and keep warm.

In a small bowl, whisk together water and flour. Place the roasting pan with the marinade on top of the stove. Over medium-high heat, whisk the flour mixture into the marinade, then add the remaining 2 teaspoons ginger. Simmer, stirring constantly, until thickened.

Slice the beef and transfer to a serving platter. Pour the hot gravy over the beef and serve.

YIELD: 10 to 12 servings

MEXICAN LASAGNA

A new twist to an old favorite. This "lasagna" will freeze for 6 months.

2 pounds ground beef round, thawed if frozen
½ cup chopped onions
1 clove garlic, minced
2 fresh green chili peppers, diced
4½ teaspoons chili powder
3 cups Basic Tomato Sauce (page 56), thawed if frozen
1½ cups cooked pinto or kidney beans
vegetable oil for frying

12 corn tortillas, thawed if frozen
2 cups ricotta cheese, thawed if frozen
1 egg, beaten
2 cups shredded Monterey Jack cheese
1 cup shredded cheddar cheese
½ cup thinly sliced scallions, optional
½ cup sour cream, optional

Combine beef, onions, garlic and peppers in a large skillet. Cook over medium heat, stirring occasionally, until meat is browned. Drain fat if necessary.

Stir in chili powder. Add tomato sauce and beans. Simmer, covered, over low heat for 15 minutes.

While sauce cooks, heat about ¼ inch of oil in a skillet. Fry tortillas, one at a time, for about 30 seconds on each side. Drain on paper towels or brown grocery bags.

In a small bowl, mix together ricotta cheese and egg. ·

Spread one-third of the meat mixture on the bottom of two 8 × 8-inch ovenproof casseroles or one 9 × 13-inch baking pan. Cover with half of the Monterey Jack cheese, half of the ricotta-egg mixture and six tortillas. Break tortillas if necessary to cover cheese more completely. Repeat using last third of meat mixture as the final layer. At this point the casserole may be covered and frozen for later use.

When ready to bake, top with cheddar cheese and bake, uncovered, in a 350°F oven for 45 to 50 minutes or until hot and bubbly. If frozen, bake, covered, at 350°F for 1 hour, then uncover and bake 30 minutes more. Serve with scallions and sour cream if desired.

YIELD: 10 to 12 servings

CHILI CON CARNE
WITH KIDNEY BEANS

If you make this chili with fresh meat, you can freeze it for 6 months. Freeze it for 3 months if made with frozen meat. It's okay to refreeze chili made with frozen vegetables.

2 tablespoons corn oil
1 pound lean ground beef,
   thawed if frozen
2 cloves garlic, minced
2 tablespoons chili powder
2 cups frozen chopped
   tomatoes
1 cup cooked kidney beans

2 tablespoons whole grain
   yellow cornmeal
½ cup shredded sharp
   cheddar cheese
1 tablespoon finely
   chopped fresh
   coriander leaves or
   parsley

Heat oil in a heavy 3- to 4-quart Dutch oven over high heat. Break off small pieces of ground meat and add quickly, stirring with a wooden spoon until meat loses some of its red color.

Lower heat; add garlic, chili powder and tomatoes. Cover pot, lower heat and simmer for 15 minutes. Add beans and cook 10 minutes more. Sprinkle with cornmeal and stir. Cook for 5 minutes to thicken.

Cool and freeze at this point or, if preparing immediately, sprinkle with cheese and cover until cheese is melted, about 5 minutes. Sprinkle with coriander or parsley.

YIELD: 2 to 3 servings

MOROCCAN LAMB
AND YELLOW SQUASH STEW

Serve this spicy North African lamb stew over a mound of cooked millet.
Make a large quantity and freeze all but one dinner's portion. It can be
frozen for 6 months.

3 pounds lean lamb,
   partially thawed if
   frozen
½ cup butter
2 large onions, finely
   chopped
3 cloves garlic, minced
4 tablespoons whole wheat
   flour
2½ cups Chicken Stock
   (page 73) or Beef Stock
   (page 71), thawed if
   frozen
½ teaspoon saffron threads
¾ teaspoon ground
   cardamom

¾ teaspoon grated fresh
   gingerroot, or
   ¼ teaspoon ground
   ginger
¼ teaspoon ground allspice
1 teaspoon ground
   cinnamon
1⅓ cups golden raisins
   pinch of cayenne pepper
6 cups sliced yellow
   summer squash,
   partially thawed if
   frozen
   juice and grated peel of
   2 lemons

Slice lamb into ½- to 1-inch strips. Heat butter in a Dutch oven and brown
the lamb on all sides. Remove the lamb with a slotted spoon and set aside.

In the Dutch oven, sauté onions and garlic until onions are tender.
Mix in flour and cook, stirring constantly, until flour is toasted a golden
brown. Add stock, stirring constantly to avoid lumps. Then add the meat,
saffron, cardamom, gingerroot, allspice, cinnamon, raisins and cayenne.

Cover and simmer over low heat, stirring occasionally, for 1 hour, or
until meat is tender. You may want to freeze the mixture at this point, or
add squash and cook for about 10 minutes more. Remove from heat and
stir in lemon juice and peel.

To serve stew that has been frozen, reheat in a large saucepan over
low heat. When thoroughly heated, add squash and cook 10 minutes
longer, then remove from heat and add lemon juice and peel.

YIELD: 8 to 10 servings, 2 to 2½ quarts

VARIATION: Winter squash can be used in place of summer squash.

CASSOULET

Cassoulet, a hearty French casserole, is traditionally made with goose or duck. Leftover chicken or turkey is an excellent substitute. Since this recipe makes a large quantity, half may be frozen for later use. It will freeze for 6 months.

2 pounds dried white beans, soaked overnight (or see Beans, Dried, page 204)	3 cups Beef Stock (page 71), thawed if frozen
1 pound pork rind, cut into small squares	freshly ground black pepper, to taste
10 full sprigs fresh parsley	2 teaspoons soy sauce, optional
6-8 sprigs fresh thyme, or 2 teaspoons dried thyme	1½ cups water
2 cups diced onions	1 pound sausage (garlic, Italian or kielbasa), thawed if frozen
6 cloves garlic, finely chopped	1 pound cooked (roasted if possible) poultry, cut into serving-size pieces with bones intact, thawed if frozen
2 bay leaves	
3 whole cloves	
2 tablespoons olive oil	
1 pound lamb, mutton or pork, cut in chunks, thawed if frozen	1½ cups bread crumbs
2 cups chopped onions	1 cup chopped fresh parsley
¾ cup tomato paste	

Place drained beans in an 8-quart pot and cover with water.

Place pork rind in a piece of cheesecloth and tie securely so that the pork rind can be removed from the pot easily.

If using fresh parsley and thyme, tie the sprigs (use only three or four of thyme) together with string.

Add pork rind, parsley, thyme, diced onions, half of the garlic, bay leaves and cloves to the beans and bring to a boil, stirring occassionally, then lower heat to a simmer. Skim off and discard any foam that forms on the surface. Do not allow the water level to fall below the level of the beans. Add boiling water if it is necessary to add water.

When beans are cooked, about 45 minutes to 1 hour, set the pot aside and let the beans soak in the herbal broth.

Meanwhile, heat oil in a medium-size saucepan. Sauté meat chunks until well browned. Add chopped onions and cook on low heat until onions are translucent. Add remaining thyme and garlic, tomato paste, stock, pepper and soy sauce, if desired. Bring to a boil, then lower heat to simmer.

While meat chunks are simmering, bring water to a boil in a medium-size saucepan. Prick the sausage in several places and cook in the water until the meat has lost its pink color, 20 to 30 minutes. Pour off any remaining liquid and brown the sausage well. Drain on paper towels and set aside. Cool and cut into 2-inch pieces.

When the meat chunks have simmered for an hour or so, remove them from the pot and set aside.

Drain the beans, removing the pork rind, parsley, thyme, bay leaves and cloves. Reserve broth.

Add the beans to the meat broth and bring to a boil. Remove from heat and set aside for 15 minutes. Divide the beans, meats and broth equally between two 2-quart ovenproof casseroles. (Earthernware is best for immediate use. Use glass or metal for frozen portion.) Start by spooning one-sixth of the beans into each of the casseroles. On top of the beans arrange a layer of each meat, using one-fourth of the total in each casserole. Spoon over another sixth of the beans, followed by the remaining meat. Add the remaining beans. Pour enough of the reserved bean broth into the casseroles so the liquid comes up to the top of the beans.

Make a mixture of the bread crumbs and chopped parsley and sprinkle a thick layer over the beans.

Bake cassoulets uncovered in a 350°F oven. If serving immediately, bake for 2 hours. Every 30 minutes check cassoulets. If the tops are crusty, break the crust and spoon a bit of liquid from the casseroles over the top. Repeat this for the duration of the cooking time. The tops should be crusty and well browned, but casseroles should not be dried out deeper down. Add more reserved bean broth if necessary. Remove cassoulet for freezing after 1½ hours. (The remaining cassoulet requires 30 minutes additional baking time.)

To bake a frozen cassoulet, place directly in a 350°F oven and bake until hot, about 1 hour, then bake an additional 30 minutes.

YIELD: about 16 servings, 4 quarts

PORK CHOPS WITH CARAWAY STUFFING
AND SOUR CREAM GRAVY

These stuffed chops can be frozen for 3 months, or the stuffing can be frozen by itself for 3 months.

4 1-inch-thick loin pork chops with pockets, thawed if frozen ground black pepper, to taste	1 cup rye bread crumbs
	½–1 teaspoon caraway seeds
	3 tablespoons chopped fresh parsley
2 tablespoons butter	1 cup water
1 medium-size onion, chopped	2 tablespoons whole wheat flour
1 clove garlic, minced	½ cup sour cream

Season the pork chops inside and out with pepper. Set aside.

Melt the butter in a small skillet. Add the onions and garlic. Sauté until onion is translucent, about 5 minutes.

In a small bowl, mix onions, garlic, bread crumbs, caraway seeds and parsley together well. At this point, the stuffing may be frozen for later use. Completely defrost filling before stuffing chops.

Stuff the bread mixture into pork pockets. At this point you may freeze the stuffed chops. If baking, place ¼ cup water and the chops in a baking pan. Cover and bake at 350°F for 30 minutes. Uncover and bake until brown and tender, about 30 minutes more.

Remove chops to a serving platter and keep warm. Pour drippings into a small saucepan. Make a paste from 1 tablespoon water and flour. Whisk paste into drippings. Add remaining water. Cook over medium heat, whisking constantly, until thickened. Turn heat to low. Whisk in sour cream and heat gently. Add pepper to taste and serve with the stuffed pork chops.

To make frozen stuffed pork chops, thaw overnight in the refrigerator, then bake as directed.

YIELD: 4 servings

BARBECUED SPARERIBS

Can be frozen for 4 months if made with fresh ribs.

1 medium-size onion,
finely chopped
2 cloves garlic, minced
4 tablespoons butter
¾ cup tomato paste
2 tablespoons honey
2 tablespoons Dijon-style
mustard
½ cup apple cider vinegar

¼ cup water
½ teaspoon freshly ground
black pepper
1½ teaspoons Tabasco sauce
(or more to taste)
1 · tablespoon soy sauce
4 pounds meaty pork
spareribs, thawed if
frozen

Sauté onions and garlic in butter until soft but not browned. Add tomato paste, honey, mustard, vinegar, water, pepper, Tabasco sauce and soy sauce, stirring to blend. Simmer over medium heat for 15 minutes, stirring often. Taste and adjust seasoning. This sauce can be frozen now for later use.

Preheat oven to 450°F. Place spareribs with meaty side up in large baking pan and roast for 20 minutes. Remove pan from oven and pour off any fat that has accumulated. At this point, ribs can be frozen with sauce in aluminum foil. Or turn oven temperature down to 350°F, return the ribs to the pan, spread sauce over them and place ribs in the oven and bake for 1 hour.

Reheat frozen ribs at 350°F just until hot, about 1 hour and 20 minutes, in the same foil in which you froze them.

YIELD: 4 servings

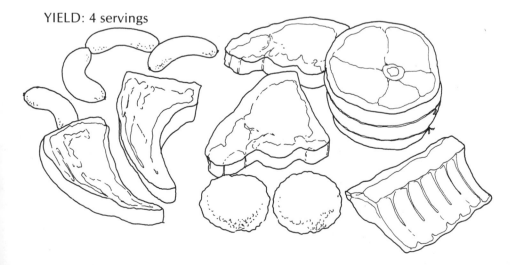

INDONESIAN SATE

This is the Indonesian version of shish kebab. It is pronounced *sah*-tay. The meat marinates in the freezer.

¼ cup vegetable oil
1 cup chopped onions
1 tablespoon minced
   fresh gingerroot
2 tablespoons soy sauce
2 tablespoons lemon juice
2 cloves garlic, minced
¼ cup honey
1 teaspoon ground black
   pepper

½ teaspoon ground
   turmeric
½ teaspoon ground cumin
1 pound pork loin, cut into
   1-inch cubes
1 pound pork liver, cut
   into 1-inch cubes
1 pound boned and skinned
   chicken breast, cut
   into 1-inch cubes

Place a doubled, large freezer bag in a large bowl. Add oil, onions, gingerroot, soy sauce, lemon juice, garlic, honey, pepper, turmeric and cumin. Knead and shake gently until mixed. Add pork loin, pork liver and chicken. Mix again. Freeze for 3 to 5 days.

Thaw in the refrigerator overnight. Thread meats alternately onto 6 to 8 skewers. Broil until meat is done, about 15 minutes. Turn and brush occasionally with marinade while broiling. Serve hot on a bed of brown rice.

YIELD: 6 to 8 servings

RED FLANNEL HASH

Making this hash is an excellent way to use leftover roast pork. While not traditional, turkey or beef would be equally good. Leftover Red Flannel Hash is delicious tossed into a green salad or marinated in oil and vinegar and topped with a sprinkle of parsley.

1 onion, minced
1 clove garlic, minced
1 tablespoon vegetable oil
3 unpeeled boiled
   potatoes, diced
2 cups frozen diced cooked
   pork, partially thawed

3 cups frozen cubed
   cooked red beets,
   partially thawed
½ teaspoon vinegar
   pinch of cayenne
   pepper

In a large skillet, cook onions and garlic in oil over medium heat until limp. Add potatoes and cook until they begin to brown lightly, adding more oil if necessary. Add pork, beets, vinegar and cayenne and continue cooking for about 5 minutes.

YIELD: 4 servings

SOUVLAKIA

The uncooked meat can be frozen in the marinade for 1 week.

1½ pounds veal, partially thawed if frozen	1 tablespoon dried oregano
¼ cup olive oil	¼ teaspoon ground black pepper
¼ cup apple juice	2 cloves garlic, minced
¼ cup lemon juice	1 small onion, chopped
¼ cup white wine vinegar	

Cut the veal into 1½-inch cubes.

In a medium-size bowl, combine the oil, apple juice, lemon juice, vinegar, oregano, pepper, garlic and onions.

Toss the veal with the marinade. Freeze it now for later use, or cover and place in refrigerator for at least 12 hours. Toss the meat occasionally.

Remove meat from marinade with a slotted spoon and thread onto skewers. Place the skewers on a broiler pan, place the pan on the lowest oven rack and broil for 10 minutes. Turn the meat and baste it with marinade. Broil for 10 minutes more. Repeat turning, basting and broiling until all sides are lightly browned. Veal can be served on or off the skewers.

To use frozen meat, thaw completely in refrigerator before threading onto skewers. Broil as directed above.

YIELD: 4 to 6 servings

VARIATIONS: Use lamb, pork, beef or fish steaks in place of veal.

LEEK AND SAUSAGE PIE

The filling for this pie can be frozen for 3 months.

½ pound sausage, thawed
   if frozen
2 tablespoons butter
2 cups sliced leeks,
   thawed if frozen
1 cup sliced mushrooms
1 tablespoon Mochiko
   rice flour
¾ cup light cream or milk
1 egg, lightly beaten
1 tablespoon minced fresh
   parsley

freshly ground black
   pepper, to taste
½ teaspoon ground nutmeg
1 tablespoon finely
   chopped fresh dillweed
1 baked 9-inch whole
   wheat pie shell
   (page 329), thawed
   if frozen
¼ cup grated Parmesan
   cheese

Cut sausage into bite-size pieces. In large skillet, sauté sausage in 1 teaspoon of the butter until just cooked. Drain excess fat and reserve meat. Sauté leeks and mushrooms in remaining butter until leeks are translucent.

Combine rice flour and 3 tablespoons of the cream to make a paste. Stir in remaining cream and egg. Over low heat, add flour mixture and sausage to the leeks and mushrooms. Cook, stirring, until slightly thickened.

Turn off heat. Add parsley, black pepper, nutmeg and dillweed. Stir to combine ingredients. Freeze the filling for later use or pour into pre-baked crust.

To serve, bake 20 minutes in a 375°F oven. Sprinkle cheese over the pie and bake another 10 to 15 minutes, or until nicely browned.

To prepare pie with frozen filling, thaw in refrigerator overnight, then pour into prebaked pie shell and bake as above.

YIELD: one 9-inch pie

VARIATION: For a lighter pie, replace the sausage with ½ pound of steamed asparagus.

GREEK MACARONI AND MEAT CASSEROLE

You may wish to prepare this in two 1½-quart casseroles—one to bake right away and one to freeze for up to 3 months.

⅓ cup finely chopped onions	½ teaspoon dried thyme
1 tablespoon vegetable oil	1 cup Basic Tomato Sauce (page 56), thawed if frozen
2 pounds ground beef, pork or veal or combi-nation, thawed if frozen	½ cup melted butter
1 teaspoon soy sauce	1 cup grated Parmesan cheese
¼ teaspoon ground black pepper	2 cups whole wheat elbow macaroni, cooked
2 teaspoons chopped fresh basil, or 1 tea-spoon dried basil	¼ cup Mochiko rice flour
	3 cups milk
	4 eggs

In a large skillet, sauté onions in oil until soft. Add meat and cook over medium heat until brown. Drain all fat from meat. Stir in soy sauce, pepper, basil, thyme and tomato sauce. Cover and simmer for 25 minutes. Set aside.

Combine 2 tablespoons of the melted butter and ¾ cup of the cheese in a medium-size bowl. Add macaroni and mix well. Set aside.

Heat remaining butter in a 2-quart saucepan over medium heat. Whisk in flour. Add milk slowly, while stirring to prevent lumps. Cook, stirring constantly, until mixture forms a thick, smooth white sauce.

Remove from heat. In a medium-size bowl, beat the eggs lightly. Add 2 cups of the white sauce very slowly and blend well. Pour this mixture over macaroni and stir gently to coat all the macaroni.

Put half of the macaroni mixture into a 3-quart buttered overproof casserole or divide into two 1½-quart casseroles. Layer with half of the meat mixture. Repeat. Top with the reserved white sauce and sprinkle with remaining cheese.

At this point the casserole may be frozen.

Bake at 400°F for 40 to 45 minutes or until browned and bubbly. If frozen, bake covered at 350°F for 1 hour and 15 minutes to 1½ hours. Uncover during last 15 minutes to allow the top to brown.

YIELD: 8 to 10 servings

CABBAGE STUFFED WITH GROUND MEAT

Stuffed cabbage is a popular peasant food prepared, with many variations, all over Eastern Europe. A slice of good rye bread rounds out this meal. Since only the outside cabbage leaves are used, consider taking leaves from cabbage heads throughout the year and freezing them for "stuffers" as needed (see page 207 for directions on freezing cabbage leaves). This dish can be frozen for 6 months if made with fresh meat, 3 months if made with frozen meat.

1 large head cabbage, or 25 frozen cabbage leaves, thawed
1 pound of any combination of bulk sausage, ground beef or ground pork, thawed if frozen
½ cup finely chopped onions
2 cloves garlic, minced

1 cup slightly undercooked brown rice
2 eggs, lightly beaten
¼ teaspoon ground black pepper
1 cup Beef Stock (page 71), thawed if frozen
4 cups tomato juice
¼-½ cup tomato paste, optional

If using a head of cabbage, steam it for 8 minutes. Cut around core and gently pull off 25 leaves. You may have to steam the cabbage further to remove enough leaves. Any extra leaves can be frozen at this point.

Gently combine meat, onions, garlic, rice, eggs and pepper. Place ¼ cup of this meat stuffing on the face side of a leaf, 1 inch away from the bottom of the leaf. Fold bottom end over filling, then fold leaf sides in toward the center and roll tightly toward the tip.

Place stuffed leaves in a double layer, flap side down, in a large Dutch oven. Combine stock and tomato juice and pour over cabbage rolls, just to cover.

Cover and simmer gently for 1½ hours. Shake pan occasionally but do not stir. When finished, cabbage should be very tender and meat thoroughly cooked. If a thicker sauce is desired, remove cabbage rolls to a serving dish and thicken sauce with tomato paste.

Freeze now or serve in large shallow bowls with sauce.

Thaw in refrigerator overnight before reheating. Cover and cook over low heat until heated through.

YIELD: 4 to 6 servings

VARIATION: A sweet and sour sauce can be made by adding ½ cup red wine vinegar and ⅓ cup honey to the stock and tomato juice.

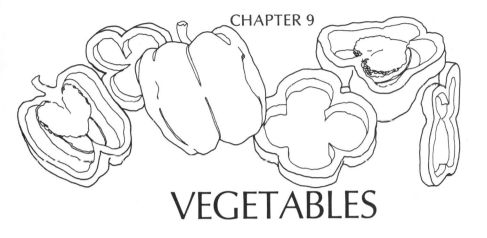

CHAPTER 9

VEGETABLES

Every summer the challenge reasserts itself. Gardens and market stalls are bursting at the seams with ripe and succulent produce. The task of preserving that bounty is at hand, and it will clearly take some dedication, and more than a little hard work. The rewards are worth the effort, that's clear. It's also clear that the job is only worth doing if it's done well and done gladly. For gardeners, particularly, preservation is happy work because it guarantees many months of healthful eating.

It starts with vegetables that are absolutely top-quality. If you freeze vegetables when they are immature, they will be soft and shriveled when you cook them. If they are overripe when you freeze them, they will be stringy and tough when cooked. Buy or pick for the freezer only those vegetables that are at their peak of freshness. It is precisely that freshness, after all, that you are trying to preserve.

No matter what the quantity of produce you undertake to freeze, handle it with speed and with respect. It was a long time growing, and the last moments it spends in your care on the way to the freezer can make a significant different in its food value.

You can, for instance, double the amount of vitamin C in the vegetables you freeze if you understand a little about harvesting. Wait to harvest your green vegetables and tomatoes until there has been a succession of several sunny days. Vitamin C fluctuates, fragile vitamin that it is, with the intensity of the light that the plant receives. Harvest in a sunny period, and while the sun is still shining, and you will have given your vegetables the maximum opportunity to produce vitamin C.

Have all the paraphernalia at hand for blanching and freezing before you bring your crops into the kitchen. That includes rounding up as many willing hands as possible to help you. Clear your calendar for the morning or the afternoon, so you can get the vegetables into the freezer within three hours of harvesting. Quality just doesn't keep. The time your vegetables spend on the countertop or in the refrigerator while you organize your resources for freezing will diminish nutritional value.

Prompt processing helps retain vitamin C, and it happens that many of the vegetables that are high in vitamin C are the very vegetables that freeze well. They include broccoli, sweet peppers, brussels sprouts, cabbage and kohlrabi.

Blanching

Blanching is actually parboiling, and no one wants to do it. It makes the kitchen hot and it slows down the passage of the produce to the freezer. It's work when you'd rather be playing. But for most vegetables blanching is just not an optional procedure. In all but a few cases, it's a necessity. Why? Enzymes are at work in your vegetables. You've read about them before in several other chapters in this book. It's the enzymes that break down the vitamin C, and convert the sugars into starches as soon as you harvest corn and peas. It's enzymes that destroy the flavor and texture in vegetables. Freezing slows them down, but it doesn't stop their action; however, heat does. And that's what blanching is. It's the brief application of high heat, either by steam or boiling water, that inactivates the enzymes and sets the color and flavor.

You may not believe that blanching is necessary. You may feel inclined to go ahead and freeze your vegetables without blanching them. In fact, we've tried it, too. The staff at the Rodale Test Kitchen froze batches of blanched and unblanched cut green beans, cut corn, corn on the cob and broccoli spears.

Three months later, they took the vegetables out of the freezer and reheated them. There was just no contest. The blanched vegetables all had bright color, sweet, fresh flavor and good texture. The unblanched vegetables all had dull color and poor flavor. The unblanched broccoli was especially mushy, bitter and rather smelly. The beans tasted bad and had a tough, fibrous texture. The cut corn managed to be both chewy and mushy at the same time, and the corn on the cob exceeded itself by being both bitter and bland.

You can be assured that we don't suggest blanching out of some purist intentions. It's really necessary, and not just to make vegetables look and taste better. They also retain more of their vitamin content when blanched, even when you allow for vitamin loss in the blanching process itself. The longer that vegetables are stored the more noticeable the contrast in vitamin content is. Blanched green beans, to cite just a single example, were found to have 1,300 percent more vitamin C than unblanched green beans after nine months. Tests have also reached similar conclusions for vitamins B_1 and B_2 and for carotene.

If you think of blanching as a way to keep in contact with your food when it is at its most perfect moment, a little of the burden is removed from the task. Get your fill of the freshness of the vegetables before you consign them to bags and containers. It's well worthwhile to put in the time handling the food you are going to freeze, because once you freeze it, it loses its tactile quality and its aroma. Create a memory of the food at its peak of flavor and freshness and you will retain enthusiasm for what you have frozen. It's also true that what you prepare now is work you don't have to do later.

Do Some Advance Planning

We hardly need tell you that you shouldn't plant more vegetables than you're willing to process or to eat. Aim to use all your freezer produce up within a year, or even less, because some vegetables don't really have a year in the freezer before they're coming up again in the garden. You will surely want to eat new vegetables from the garden in preference to those still left in the freezer from the previous summer. In fact, although homegrown vegetables will keep in the freezer for twelve months (commercially frozen vegetables for fourteen to sixteen months), plan to freeze only enough vegetables to carry you through about nine months. You don't have to freeze everything you grow. If you're freezing more of one vegetable than you can expect to willingly consume in a year, you're overdoing it.

When you bring your vegetables into the kitchen, rinse them quickly in cold running water. Soaking can leach out water-soluble B and C vitamins. If you find it necessary to peel root vegetables, do it as thinly as possible to minimize loss of the vitamins and minerals that lie just under the skin. Cut vegetables into sizes that you can most readily use for cooking.

A volunteer assembly line, lavishly fed as a reward, can be divided into people who wash, peel or cut the vegetables, or weigh them, blanch them, cool them or bag and freeze them.

Blanching Methods

Both steam and water blanching have their advocates. At the Rodale Test Kitchen, steam blanching is the favorite. Steam blanching preserves more of the nutrients, the flavor and the color of vegetables. Because the vegetables do not touch the water, fewer nutrients have the opportunity to leach out with steam blanching.

Water blanching, on the other hand, is better at removing bacteria, yeasts, molds, residues and foreign flavors. It can float out cabbageworms from broccoli and cauliflower that steam blanching can't reach. It's also marginally easier and a little bit quicker. It takes steam about 50 percent longer to penetrate to the center of the food portion. Steam blanching requires slightly more scrutiny. You have to steam the vegetables in a thin layer and make sure about halfway through the process that all portions of the vegetables are getting the same even heat. Nevertheless, at least for delicate vegetables like spinach, cabbage leaves, broccoli florets and asparagus tips, it's better.

The merits of blanching vegetables in a microwave oven are still the subject of debate. Thus far, there are only a few comparative studies to draw on. Generally, the studies indicate that steam and water blanching both produce superior results to microwave blanching. Tests conducted at Washington State University employed each blanching method, followed by freezing and thawing of the produce. The results, published in 1981, reported superior color and flavor, as well as generally higher levels of ascorbic acid retention, after steam and water blanching. These latter two methods are the ones we choose to use at the Rodale Test Kitchen.

Royal Burgundy bush beans and Royal Purple cauliflower are very cooperative, as vegetables go. They both indicate when they've been blanched long enough by changing their purple color to green. It's a built-in green light signaling "take me to your freezer."

How to Steam Blanch

First prepare the vegetables. Add 1 or 2 inches of water to the pot or blancher. When it boils, insert the steamer or basket with one layer of food on the bottom. The basket should hold the vegetables 1 or 2 inches above the water. Put the lid on. When steam starts escaping from under the pot lid, start timing the blanching. Check halfway through the blanching time that the vegetables are not clumping together and heating unevenly.

If you use two steamer baskets, you can have one filled and ready to go into the boiling water as soon as you lift the other one out of the boiling water and plunge it into ice water. Don't use a small steamer. Since it can hold only small quantities, you may be tempted to fill it up only to have the contents tumble out as soon as you try to remove them.

How to Water Blanch

Prepare the vegetables, then divide them into groups weighing about 1 pound each. Boil a gallon of water in a blancher or a big kettle. Place the vegetables in either a mesh basket or blanching basket, immerse them in the boiling water and cover. (Some people use an old deep-fat fryer.) When the water returns to a boil, start timing. Precise timing is important. There is some controversy concerning whether timing begins when you add the vegetables to the water or whether it begins when the water returns to a boil.

If you measure 1 pound of vegetables into 1 gallon of boiling water, the water will not stop boiling, so you can begin timing as soon as you add the vegetables. Not everyone does it that way, however. Some just fill their favorite container with water, then fill the blanching basket and add it without any prior measuring of ingredients or water volume. If they then start timing from the moment when the vegetables hit the water, the water may take a considerable time to return to a boil, and timing minutes may run out before the center of the food is hot enough. Underblanching will provide enough heat to actually accelerate the activity of the enzymes whose activity you are trying to slow down in the first place by blanching. Therefore, it's a good idea to stick with the pound of vegetables to a gallon of water ratio. If you overblanch your vegetables you can compensate by heating the vegetables for a marginally shorter time when you remove them from the freezer.

When steaming vegetables, it's best to do about 1 pound at a time to keep the water from losing its boil when lowering in the vegetables.

The Ice Water Plunge

If you don't cool down your vegetables as soon as they emerge from the boiling water or the steam, they'll get soggy. Allow the same amount of time for cooling your vegetables as you did to blanch them. Some cooks immerse the vegetables directly into a sink filled with icy water. Others hold them under running water. Others don't soak the vegetables, even briefly, because of the nutrient loss and instead spread them out on a cookie sheet or in a large bowl and set the bowl over ice. Others fill a large roaster with cold water and ice and set a strainer in it to hold the vegetables as they cool. Where does the ice come from for these procedures? For a lot of people, it comes from the half-gallon milk cartons of water they've been tucking into their freezers for several months ahead of time, with blanching in mind.

Once cooled, make sure you dry the vegetables before packaging them or else the water will form ice in the packages. You can dry the vegetables on paper towels spread out on the counter, but if you spread cloth towels under the paper towels you will need fewer of the latter. Some people set a table fan about 5 inches from the vegetables to speed cooling. Others give vegetables a quick ride in a salad spinner to dry them right after the ice bath.

Packing Vegetables
for the Freezer

In tests at the Rodale Test Kitchen to determine the best ways to package vegetables for the freezer, the staff worked with that old garden standby, green beans. The beans were steam blanched, then cooled in ice water. Some were tray-frozen before being added to the freezer container. Some were frozen in plastic freezer bags. Some were frozen in plastic freezer boxes, and some were sealed with a vacuum sealing machine. In six months all the packages came out of the freezer for testing.

Eager eaters decided that all the methods had worked well. The fast freezing provided by tray-freezing and the maximum removal of air provided by vacuum packing did not make a significant difference. Therefore, pack your vegetables the easiest ways. Skip tray-freezing and vacuum packing, since they create more work than they are worth.

Tray-freezing is advised as a good technique to keep vegetables separate in the freezer, but if you knock frozen vegetables hard to loosen them, you can achieve the same end. Of if you shape the bags or containers once an hour for the first three or four hours your vegetables are in the freezer, you can keep them loose. (Tray-freezing is still a good method for keeping frozen fruit loose, however, because fruit is too juicy and delicate to knock around so much.)

If you put your vegetables in containers, you needn't allow for headspace because there is no liquid that needs room to expand. If you pack vegetables in bags, use a stand to hold the bags open and a wide-neck funnel to help in pouring in the vegetables and keeping the bags clean. Suck the air out of freezer bags with a straw before you close them.

When you put the vegetables in the freezer, spread the packages throughout the freezer to promote fast freezing. The faster your vegetables freeze, the smaller the ice crystals that will form and the better the texture your vegetables will have.

Thawing and Cooking Vegetables

In many cases, freezing makes vegetables too soft to serve raw. Except for corn-on-the-cob, leafy greens and tomatoes, vegetables should go straight from the freezer to the steamer. Our green bean tests showed that beans that thawed as they cooked were considerably sweeter and firmer than beans thawed either in the refrigerator or under running water.

If you steam the vegetables (the best method), use a minimum of water: ½ cup is sufficient for an average-size pot. If the vegetables are frozen into a solid lump, break the lump into chunks by giving the bag a few hard knocks before you cook it so the heat will penetrate more evenly. Since the vegetables have already been blanched, steaming time is reduced to a third of that for fresh vegetables. You should always cook vegetables as briefly as possible in any case. All the benefits of careful blanching will be lost if you allow the reheating process to go so long that it dulls the flavor, odor and texture of the vegetables.

Stir-frying frozen vegetables also works. It's especially good for the thicker, slower-cooking vegetables like broccoli and cauliflower, which are often blanched expressly to ready them for stir-frying. When you add the vegetables to soups or stews, another good use for them, do it very near the end of the cooking process; otherwise, the vegetables will lose their texture with overcooking.

Wet Packing Vegetables

When vegetables are sautéed in oil or butter, rather than being blanched before freezing, they are called wet packs. This technique works for mushrooms,

When you freeze vegetables, follow this rule of thumb. The vegetables you normally cook will freeze well. The vegetables you usually eat raw will probably not freeze well.

summer squash, eggplants and tomatoes. If you shred root vegetables, including turnips, parsnips and rutabagas, then sauté them until they are just soft, you can freeze them successfully. Thaw the vegetables in a buttered pan in a 350°F oven. Add cream after 20 minutes, top with bread crumbs and cook until browned, another 30 minutes.

Pureed Vegetables

Good vegetable purees are not just mushy food for the baby. They have lately become one of the cornerstones of the new emphasis in the United States on eating light and eating healthy. Not only do vegetable purees, at their best, have texture and character, they also, with the exception of celery puree, freeze beautifully. They take up very little freezer space and last for a year.

Purees, in their simplicity of taste and texture, are a perfect complement for hearty roasted meats or birds. You can use them to make dips simply by mixing them with sour cream and herbs. A platter of fresh vegetables with several bowls of such dips made from vegetables of different colors, such as tomatoes, beets, carrots and zucchini, is a stunning presentation.

Hot purees, enriched with onion and cream or whipped with egg yolks and a little oil and seasonings, make highly colored accompaniments to fish dishes.

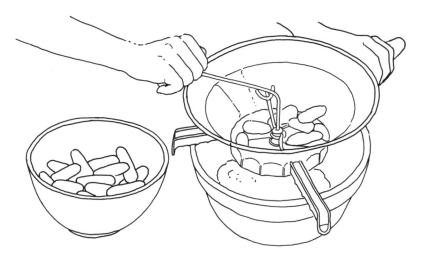

Food mills are great for pureeing tough and fibrous fruits and vegetables such as pineapples, carrots, celery and fennel and for removing skins and seeds of apples and pears, tomatoes, eggplant and green beans.

Softer fruits and vegetables such as strawberries, raspberries, bananas and squash
can be pureed by pushing them through a sieve with a rubber scraper. A drum
sieve, left, or stainless steel sieve, right, both work well.

They can also be the basis for soups and soufflés and sauces. And they can be right
on hand in your freezer.

To make a puree, start by steaming or boiling the vegetable until it is tender
enough to mash. In order for the puree to be thick and firm, you must first drain
the vegetables well after they are cooked; otherwise, the residual water combined
with the natural water content of the vegetables will create a watery puree. Reserve
the drained liquid and use it in a soup or stock.

Any vegetable can be pureed in a food processor, blender or food mill
initially, but it may well need further refining after that in order to remove skin,
seeds and fibers. Lentils, green beans, tomatoes, celery, celeriac, fennel and
eggplants will all need further processing.

If you begin pureeing with a food processor, run it in short on/off turns and
stop often enough to scrape the sides with a spatula to ensure that all the vegetables are being processed evenly. Starchy vegetables like potatoes, squash and
peas will turn pasty when pureed in a food processor and are better mashed with
a potato masher or in a food mill. A blender will also have to be watched closely
if you use it to puree vegetables, because it churns food very fine very quickly.

A food mill, on the other hand, gives you more flexibility with its choice of
fine, medium and coarse pureeing disks. Vegetables with firm textures, like
cauliflower, can be run easily through a food mill.

For the final straining of a vegetable that contains skin, seeds or fibers, use a
sieve. There are many kinds of sieves, but the best for pureeing is the old-fashioned,
big drum sieve. If your grandmother did not bequeath one to you and you cannot
find one, use a shallow stainless steel sieve instead (the kind with a hook at one
end to steady it on the top of a bowl or saucepan).

Push the food through the sieve with a plastic scraper or the back of a soup spoon, working one spoonful at a time. Freeze the puree. Before using it, thaw it 24 hours in the refrigerator, or 6 hours at room temperature. After you reheat the puree, if you reheat it, add a little butter just before serving to make the puree shiny and smoother still.

Preparing Vegetables for the Freezer

Artichokes

Artichokes freeze well, particularly small artichokes and artichoke hearts. To prepare them, wash and remove the outer leaves. Cut off the bottom end of the stem and trim off the top ¼ inch of the bud to remove the thorns. Steam blanch for eight to ten minutes, then plunge into ice water. Drain thoroughly and dry pack in plastic freezer bags.

Asparagus

Asparagus freezes very well. Wash the spears and trim the ends. If spears are large and woody, use a vegetable peeler to remove the tough outside. Sort into small, medium and large sizes. Water blanch or steam blanch small spears for two minutes, medium ones for three minutes and large spears for four minutes. Plunge into ice water, drain well and pack into containers.

Beans, Snap

Both green and wax beans freeze very well. Wash the beans and snap off or trim the ends with a knife. Leave the beans whole or cut them to the desired length. Steam blanch whole beans for four minutes, crosscuts for four minutes and julienne for three minutes. Or water blanch whole beans for three minutes, crosscuts for three minutes, julienne for two minutes. Plunge into ice water, drain thoroughly and pack. You can also tray freeze the beans after blanching, and when frozen solid, transfer them to freezer bags.

Beans, Dried

Freezing dried beans is a good way to keep them for long periods if you have a large quantity and do not have access to cool storage during warm weather. Freeze them just as they are. Soak before freezing to cut down on cooking time later.

VEGETABLE VARIETIES TO FREEZE

Here's a sampling of vegetable varieties that can be frozen. This is not a complete list, so if you don't find your favorites listed, experiment with a small quantity to test its freezing quality.

Asparagus
Martha Washington, Mary Washington

Beans, Snap
Bush:
Blue Lake, Daisy, Early Gallatin Stringless, Giant Stringless, Provider, Roma, Royal Burgundy, Tendercrop, Tenderette, Tendergreen, Top-Crop Stringless
Pole:
Blue Lake, Kentucky Wonder, Romano

Beans, Lima
Baby Fordhook, Burpee Bush Henderson, Fordhook, King of the Garden

Beets
Crosby, Detroit Dark Red, Early Wonder, Ruby Queen

Broccoli
Bonanza Hybrid, De Cicco, Green Comet Hybrid, Green Goliath

Brussels Sprouts
Jade Cross, Long Island Improved

Cabbage
Chinese, Copenhagen, KK Cross Hybrid, Pointed, Red Savoy, Surehead

Carrots
Gold Pak, Imperator Neck, Nantes Half-Long, Nantes Neck, Red-Cored Chantenay

Cauliflower
Burpeeana, Early White Hybrid, Purple Head, Royal Purple

Celery
Fordhook, Giant Pascal

Corn
Country Gentleman, Golden Cross Bantam, Honey and Cream, Kandy Korn, Silver Queen

Eggplant
Black Beauty, Ichiban Hybrid

Greens
beet greens, collard greens, dandelions, kale, kohlrabi, mustard greens, spinach, Swiss chard, turnip greens

Kohlrabi
Early White Vienna, Early Purple Vienna

Okra
Clemson Spineless, Lee, Park's Candelabra Branching, White Velvet, Ladyfinger

Onions
Ebenezer, White Bermuda, Sweet Spanish (Valencia)

Parsnips
Hollow Crown

Peas
Blue Bantam, Burpeeana Early, Freezonian, Little Marvel, Maestro, Perfected, Freezer 60, Prosperity, Thomas Laxton, Wondo

Peppers, Sweet and Hot
California Wonder, New Ace Hybrid, Anaheim Chili (hot), Jalapeño M (hot)

Potatoes, White
Kennebec (for fries), Norland (for baking), Russet Burbank (for baking)

Potatoes, Sweet
All Gold, Bush Porto Rico, Centennial

Squash, Summer
Early Golden Summer Crookneck, Patty Pan Hybrid, zucchini

Squash, Winter
Striped Cushaw, Waltham Butternut

Tomatoes
Big Boy, Heinz 1350, Marglobe, Roma, San Marzano

Turnips
American Purple Top

After soaking, rinse and drain the beans, then freeze in bags. When ready to use, tap the bag on the counter to separate the beans, then pour out what you need into boiling water.

Beans, Lima

Lima beans freeze very well. Discard any beans that are not perfect. Sort them into small, medium and large sizes. Water blanch small beans for two minutes, medium-size ones for three minutes and large ones for four minutes. Or steam blanch small beans for three minutes, medium-size ones for four minutes and large ones for five minutes. Plunge them into ice water, drain thoroughly and pack. After blanching, shell the beans, tray freeze them, and when frozen solid, pack them in freezer bags.

Beets

Beets are much better canned, but they can be frozen. Scrub the beets thoroughly and trim the tops, leaving about 1''-2''. Trim the root end at the base of the beet. Beets must be cooked before freezing. If they are only blanched, they become tough and rubbery when frozen. Cook the beets in water until tender (30 to 60 minutes, depending upon their size). Drain off the water, and discard it because it has a strong, earthy flavor. Skin the beets and leave them whole or slice, dice or cut them into julienne strips. Pack in freezer containers.

Broccoli

Broccoli freezes very well. Trim away all leaves and the tough, woody parts of the stem. Wash the head thoroughly to remove any insects or worms. Cut the stalks lengthwise into uniform size (tops should be 1 to 2 inches across), or section the head into florets and peel and slice the stems. Water blanch the larger pieces for four minutes, medium-size to small pieces for two to three minutes. Or steam blanch large pieces for five minutes, small or medium-size pieces for three to four minutes. Plunge into ice water, drain thoroughly and transfer to freezer containers.

Brussels Sprouts

Brussels sprouts freeze very well. Prepare them for the freezer by washing them, and trimming off any outer leaves and stem that remain. Sort them into small, medium and large sizes. Water blanch small sprouts for three minutes, medium-size ones for four minutes and large ones for five minutes. Steam blanch small sprouts for four minutes, medium-size ones for five minutes, and large ones for six minutes. Plunge into ice water, drain thoroughly and transfer to freezer bags or containers.

Cabbage

Cabbage freezes fairly well, but plan to use it in cooked dishes, not salads. Wash the head and discard any outer leaves. To remove whole leaves for freezing, first cut around the core of the cabbage to loosen the leaves. Then steam blanch the cabbage until the leaves are translucent. Cool for five minutes and remove the outer leaves. To remove the inner leaves, reblanch the cabbage. Pack the leaves flat for freezing, with a double layer of wax paper between each. You can also cut the cabbage head into wedges or shred it. Water blanch shredded cabbage for one and one-half minutes, wedges for two minutes. Or steam blanch shredded cabbage for three minutes, wedges for four minutes.

Carrots

Carrots do freeze well, but root cellar storage also works well. Wash them and peel if desired. Cut the carrots into julienne strips, large chunks, slices or, if they are small (3 to 4 inches), leave them whole. Water blanch slices or strips for two minutes and chunks or small whole carrots for three minutes. Or steam blanch slices or strips for four minutes, and large chunks or small whole carrots for five minutes. Plunge into ice water, drain well and pack.

Cauliflower

Cauliflower freezes well. Discard leaves and stem, and wash the head carefully. Break it into florets that measure approximately 2 inches across the top. Leave very small heads (about 4 inches across) whole. Water or steam blanch the florets for four minutes, whole small heads for six minutes. Plunge into ice water, drain well and pack.

Celeriac

Celeriac freezes well. Wash, peel and cut into discs or cubes. Water blanch for four minutes, plunge into ice water, drain and pack in freezer bags.

Celery

Celery freezes poorly, becoming very soft. Freeze it only to use in soups, stews and casseroles. If you do want to freeze celery, first cut off the base. Wash the stalks thoroughly, removing any tough strings or damaged spots. Cut stalks into 1-inch chunks. Water blanch for three minutes or steam blanch for four minutes. Plunge chunks into ice water, drain well and pack.

Corn

Corn freezes very well in all forms. For corn-on-the-cob, remove the husks and silk from the ears. Cut out any insect-damaged areas. Sort the ears according to size. Water blanch small ears for five minutes, medium-size ones for six minutes and large ones for eight minutes. Remove from the water with tongs and plunge the ears into ice water. Drain thoroughly, wrap each ear individually and pack them in groups in freezer bags. Thaw corn-on-the-cob in the refrigerator for nine to ten hours before reheating.

For whole kernel corn, blanch as for corn-on-the-cob, then cut the kernels off the cob as close to the cob as possible but without including any of the tough underlying cob. To steady the cobs, set them in the cone of an angel food cake pan. As you cut, the kernels will fall into the pan. Pack in bags or containers.

For cream-style corn, blanch the ears as for corn-on-the-cob. Cut the kernels off the cob by first running your knife down the middle of the kernels. Then cut off the kernel hearts. You should have a fair amount of juice when you do this. Pack in containers along with the juice, leaving some headspace.

Cucumbers

Cucumbers turn mushy when frozen, so freeze them only to use in soups, stews or casseroles. Peel the cukes and cut them into quarters, lengthwise. Remove the core and seeds and cut the quarters into chunks. Freeze the chunks or grate and freeze them or puree and freeze them. Dry pack the cucumber or bake the chunks in a 425°F oven for 30 minutes before freezing.

Eggplant

You can freeze eggplant, but it becomes very soft after freezing. Wash the fruit, peel it and cut it into ½-inch thick slices or into cubes. Dip the pieces into lemon juice. Water blanch for three minutes, or steam blanch for four minutes, then dip again into lemon juice. Drain well and pack. Place a layer of wax paper between the slices for easy removal.

You can also roast eggplant to a smoky flavor in the oven or on an outdoor barbecue, then freeze it to use later in moussaka. Or you can roast whole eggplant and freeze the pulp to use in *Baba Ghannouj* (a Middle Eastern spread) and other appetizers. To roast eggplant in the oven, prick it with a fork, then bake it on an oven rack at 400°F until soft, 40 to 65 minutes. Slice it open lengthwise and place it cut side down on a wire rack to cool and drain. Or cut fresh eggplant in half, slit the skin, place it under the broiler and broil until skin is charred and pulp is soft. Scoop out the pulp from the skins, mix in 1 tablespoon of lemon juice for each cup of pulp and freeze.

You can also sauté diced eggplant in oil along with some onion and freeze it to use in appetizers like caponata or as a pizza topping. Or bread eggplant slices and freeze them as described for green tomatoes, on page 214.

Fennel

Bulb fennel (not the herb) is an anise-flavored, celerylike vegetable that becomes very soft when frozen. Freeze it only to use in soups, stews or as a soft vegetable. Cut off the tough, upper stalks and remove the outer leaves from the bulb. Cut off the bottom end. Pull apart the sections into pieces 1½ to 2 inches long, or cut each section in half lengthwise. Steam blanch the pieces for two to three minutes. Plunge immediately into ice water and cool. Drain well and pack.

Garlic

You can freeze garlic in any of three ways. The first way is to grind or chop the garlic, wrap it tightly and freeze. To use it, just grate or break off the amount you need. Or freeze the garlic unpeeled and just remove cloves as you need them. The third method is to peel the cloves and puree them with oil in a blender or food processor, using two parts oil to one part garlic. The puree will stay soft enough in the freezer to scrape out parts to use in sautéing.

Greens

All kinds of greens—amaranth, beet greens, chicory, collards, dandelion, endive, escarole, kale, lettuce, mustard greens, sorrel, spinach, Swiss chard, turnip greens, watercress—lose their characteristic crisp texture after freezing. It's best to freeze greens only to use in soups, stews and casseroles. Don't plan on using them in salads. Wash them very thoroughly in several changes of water. Discard any damaged, wilted or insect-eaten leaves. Trim off stems and large mid-ribs. Water blanch spinach for two minutes or steam blanch for three minutes. Other greens with heavier texture may take a few minutes longer. Plunge into ice water and drain very thoroughly, then pack into bags or containers. Partially thaw greens before steaming or sautéing them.

The Rodale Test Kitchen experimented with freezing sorrel, in particular, because its season in the garden is so brief and its leaves are so delicate. The simplest effective method they found is to pour boiling water over chopped sorrel in a freezer container. Cool it and freeze. Sorrel frozen this way is useful for soups but not sauces. To freeze sorrel for use in sauces, see the recipe for Sorrel Puree, page 229 (use it in Russian Sorrel Sauce, page 53).

Kohlrabi

Kohlrabi freezes like celery, becoming very soft. Wash and trim off the trunk of the plant. Smaller vegetables can be left whole. Cut the larger ones into ½-inch cubes or slices. Water blanch cubes and slices for two minutes, whole kohlrabi for three minutes. Or steam blanch cubes and slices for three minutes, whole kohlrabi for five minutes. Gently plunge into cold water, then drain thoroughly.

Leeks

These elegant vegetables freeze well without blanching. Wash them thoroughly and cut either into chunks or long, thin slices. Put immediately into freezer bags. To separate the pieces when they are frozen, just tap the bag, then remove the amount you want.

Mushrooms

Mushrooms retain their flavor but they become softer and darker and sometimes tougher in the freezer. To freeze, wash them gently with a cloth or mushroom brush or your hands. Do not soak mushrooms, which absorb water like a sponge. Leave small, button-type mushrooms whole. Large mushrooms can be halved, quartered or sliced. If you are planning to use larger mushrooms for stuffed mushrooms, leave them whole.

Blanch mushrooms to prevent darkening. Steam blanch small, whole mushrooms for three minutes, halved, quartered or sliced ones for three minutes and large whole mushrooms for five minutes. To prevent darkening, add lemon juice (1 tablespoon to 1 quart of water) to the ice water that you use to cool the mushrooms after blanching. Drain thoroughly and tray-freeze, then transfer the mushrooms to freezer containers. Tray-freezing keeps the mushrooms separate for easy removal. An alternative to blanching is to sauté thinly sliced mushrooms in butter or oil, then freeze. Expect some shrinkage when you cook the mushrooms.

Okra

Okra freezes well. Remove the stems without cutting into the pods, and sort the pods by sizes. Water blanch small okra for one minute, medium size for two minutes and large ones for three minutes. Steam blanch small size for two minutes, medium size for three minutes and large size for four minutes.

An alternate preparation method that eliminates the gumbo mess is to slice the pods thinly, then stir-fry them in oil about three minutes to coat. If you do blanch okra, roll the pods in cornmeal after draining them, then tray-freeze them and transfer to freezer bags. When ready to prepare, they can go right from freezer bag into the frying pan without thawing.

Onions

Onions freeze well without blanching. Wash, peel and trim off any stems or roots. Chop them, or cut into slices or large pieces. Collect in a freezer bag and freeze. After freezing, the pieces will separate easily when the bag is tapped, enabling you to remove only what you need. (However, if you plan to store onions for more than three months, steam blanch for two minutes before freezing.)

For very small pearl onions, steam blanch whole onions for three to four minutes, then cool quickly in ice water. Drain thoroughly and pack.

Parsnips

Parsnips freeze fairly well. Wash, peel and slice them. Water blanch for two minutes or steam blanch for three minutes. Plunge into ice water, then drain thoroughly and pack in containers.

To cool a child's steaming bowl of soup, pour in a handful of frozen peas. They're more nutritious than an ice cube, and just as effective.

Peas

Garden, sugar snap and snow peas all freeze very well. Wash the pods, if necessary, and shell garden peas, and, if you like, snap peas. Trim the ends of snow peas and the snap peas that you will freeze in the pods. Water or steam blanch shelled peas for one and one-half to two minutes. Plunge into ice water, drain thoroughly and pack or tray-freeze, then transfer to freezer bags. Water blanch edible-podded peas for one minute or steam blanch for two minutes, plunge into ice water, drain thoroughly and pack or tray-freeze, then transfer to freezer bags.

Peppers

Both sweet and hot peppers freeze very well, either whole, sliced or in chunks. There are a couple of different ways to prepare peppers for freezing. Begin by washing them well and cutting out the stems of the sweet peppers. You can leave sweet peppers whole or slice, dice or quarter them. Leave hot peppers whole or chop them. There is no need to blanch peppers before freezing unless you plan to freeze sweet peppers whole. In that case, steam blanch whole peppers for two minutes to keep the skins from getting tough. All other forms can be tray-frozen, then transferred to plastic freezer bags.

You can also broil hot or sweet peppers, turning them until the skins are split and charred on all sides, five to ten minutes. Cool the peppers with a damp towel covering them. The steam created will loosen the skins so you can easily pull them off. Then remove cores and seeds and collect them together with the skins and any juice. Strain them and collect the juice to use in packing the peppers for the freezer.

Potatoes, White

Raw potatoes turn soft in the freezer. Freeze only cooked potatoes and reheat them straight from the freezer. That is the only way to prevent them from getting grainy and watery after freezing. The potatoes will still have a tendency to fall apart and are best used to make either potato cakes or hash brown potatoes.

To freeze potatoes, bake or steam waxy or all-purpose potatoes until they are almost cooked. Check by slipping a paring knife in, and stop the cooking when the knife still meets with a little resistance toward the center. Cut the potatoes in half to allow steam to escape as they cool. When cool, cut them into ¼-inch cubes. Freeze in a double layer in aluminum foil. They can be frozen for four months. Reheat without thawing. To make hash browns, heat oil or butter or combine them in a pan. Brown the potatoes, loosely covered, over enough heat so the oil sizzles but the potatoes cook slowly. Flip and brown the potatoes on the other side.

Potatoes, Sweet

Sweet potatoes will darken if you freeze them raw, but they freeze very well when cooked. Bake for ten minutes in the peel, freeze, then finish baking them when thawed. Or you can freeze sweet potatoes as a puree or in cooked slices. Add lemon juice to the puree and dip the slices in lemon juice before freezing to prevent darkening. Glazed sweet potatoes also freeze well, as do sweet potatoes combined with one part orange juice for every four parts mashed sweet potatoes.

Radishes

Do not freeze radishes of any sort, including Oriental types like daikon. Whether you blanch them first or not, radishes change color and become very soft in the freezer.

Soybeans

Soybeans freeze very well. Blanch them *before* shelling, though, to make shelling easier. Water blanch or steam blanch for five minutes, cool and shell. There is no need for any further blanching. Rinse the beans thoroughly in cold water, drain and freeze.

Squash, Spaghetti

Spaghetti squash freezes very well, and is handled a bit differently from other winter squashes. Wash, slit in half lengthwise and scoop out the seeds. Place squash, cut side up, in shallow baking pans and bake at 375°F for 30 minutes. Remove from the oven and cool until the squash can be easily handled. Pull a fork lengthwise through the flesh to separate it into long strands. Pack into freezer bags and freeze. To reheat frozen squash, partially thaw, then steam, stirring occasionally until tender but still firm, eight to fifteen minutes.

Squash, Summer and Winter

Winter squashes all freeze well, but they may lose their original texture. Summer squash is best suited for soup after it's been frozen, but freezing is the only way to keep it for any length of time. To freeze pumpkin and other winter squashes, wash, halve and remove seeds. You can leave the squash in halves or peel and cube it. To make a puree, bake the pumpkin and winter squash halves in a 350°F oven until soft. (Steaming is also possible but will make a wetter puree.) Scrape the meat out of the shells and mash thoroughly or run through a food mill. If preparing cubes, peel and steam the cubes until soft. Leave the cubes whole or run through a food mill or processor. Pack into containers.

To freeze zucchini and other kinds of summer squash, cut in half, remove seeds, if necessary, and cut into cubes. Steam blanch the cubes for two to three minutes. Drain well and pack the cubes in containers, or mash them or run them through a food mill before packing.

If you grate zucchini with the skin on it will stay crunchy when frozen without blanching. Use the side of the grater with the largest holes, or use a food processor with a coarse shredding plate. When you thaw the zucchini, drain it and save the juice to use in soups and breads. Add shredded zucchini to quick breads, omelets or spaghetti sauce or layered in lasagna, where it will provide some flavor but mainly texture.

Tomatoes

Frozen tomatoes keep their fabulous fresh flavor, but unfortunately, their texture becomes mushy. Freeze tomatoes only to use in soups, sauces or juice. Wash the tomatoes, remove the stems and cut out the core and any bad spots. Leave the tomatoes whole or quarter them and freeze in bags.

The skins will toughen in the freezer, but they are much easier to remove when the tomatoes are frozen. Run the frozen tomatoes under cold water and the skins will slip right off. To skin before freezing, drop the tomatoes into boiling water and let them cook for 30 seconds or just until the skins crack. Remove from the water with a slotted spoon and carefully pull off the skins. Frozen tomatoes should be thawed in the refrigerator for ten hours before you use them.

If your freezer space is limited, freeze tomatoes in the form of cooked juice, soup, puree or sauce. To make a simple tomato juice, simmer quartered tomatoes in a covered pot for about ten minutes. Run the tomatoes through a food mill to remove the skins. Cool and pour into freezer containers, leaving 1 inch of headspace. See Chapters 3 and 4 for some recipes for tomato sauces and soups.

To freeze green tomatoes, cut them into ½-inch slices, discarding the top and bottom pieces. Dust the slices with fine cracker or bread crumbs or whole wheat pastry flour. Brown in oil over medium-high heat to form a crusty coating. Work quickly so the tomatoes do not cook. Drain on cookie sheets lined with paper towels or brown paper bags. Tray-freeze the slices, collect them in freezer bags and return to the freezer. To serve the frozen slices, thaw them for twenty minutes, then sauté again. Serve with white sauce or melted cheese on top or in sandwiches with lettuce and onion.

Turnips and Rutabagas

To freeze turnips and rutabagas, cut off the tops, wash the roots thoroughly and peel them. For rutabagas, peeling is difficult. Slice off the skin as thinly as you can. Shred or cut the roots into ½-inch cubes or ¼-inch slices. Water blanch the cubes or slices for one to two minutes. Plunge into ice water, drain and pack. You can also cook the cubes until tender, then mash them or run them through a food mill. To freeze shredded turnips, sauté in butter till limp, then cool and pack.

APPROXIMATE YIELD OF
FROZEN VEGETABLES FROM FRESH

VEGETABLE	FRESH	FROZEN
Asparagus	1 crate 1 to 1½ pounds	15 to 22 pints 1 pint
Beans, Snap	1 bushel ⅔ to 1 pound	30 to 45 pints 1 pint
Beans, Lima	1 bushel 2 to 2½ pounds	12 to 16 pints 1 pint
Beets	1 bushel 1¼ to 1½ pounds	35 to 42 pints 1 pint
Broccoli	1 crate 1 pound	24 pints 1 pint
Brussels Sprouts	4 1-quart boxes 1 pound	6 pints 1 pint
Cabbage	1 to 1½ pounds	1 pint
Carrots	1 bushel 1¼ to 1½ pounds	32 to 40 pints 1 pint
Cauliflower	1⅓ pounds	1 pint
Celeriac	2 to 3 celeriac	1 pint
Celery	2 medium bunches	1 pint
Corn (in husks)	1 bushel 6 to 8 ears	14 to 17 pints 1 pint
Cucumbers	4 medium	1 pint
Eggplant	1 to 1½ pounds	1 pint
Fennel	2 stalks	1 pint
Greens	1 to 1½ pounds	1 pint
Kohlrabi	1¼ to 1½ pounds	1 pint
Leeks	3 large leeks	1 pint
Mushrooms	1 to 2 pounds	1 pint
Okra	1 pound	1 pint

(continued)

· APPROXIMATE YIELD OF
FROZEN VEGETABLES FROM FRESH *(continued)*

VEGETABLE	FRESH	FROZEN
Onions	2 to 3 onions	1 pint
Parsnips	1¼ to 1½ pounds	1 pint
Peas, Garden, Snap or Snow	1 bushel 2 to 2½ pounds	12 to 15 pints 1 pint
Peppers, Sweet or Hot	⅔ pound	1 pint
Potatoes	2 to 4 pounds	1 pint
Soybeans	2 to 2½ pounds	1 pint
Squash, Summer or Winter	1 pound·	1 pint
Tomatoes	1 pound	1 pint
Turnips and Rutabagas	1¼ to 1½ pounds	1 pint

ASPARAGUS-ORANGE MIMOSA SALAD

This is a delightful, light salad that makes use of your frozen asparagus.

¾ pound frozen asparagus	1 teaspoon grated orange
¼ cup sunflower seed oil	peel
3 tablespoons orange juice	1 head Boston lettuce
2 tablespoons lemon juice	2 hard-cooked eggs

Steam asparagus just until thawed. Cool immediately under running water. Drain thoroughly and place in a shallow bowl.

Stir together oil, orange and lemon juices and orange peel. Pour over asparagus.

Let asparagus marinate for 2 hours at room temperature. Stir, then chill briefly. Serve soon, or the asparagus will begin to lose its bright color.

Separate lettuce into leaves. Wash, dry and arrange lettuce in four shallow bowls.

Lay asparagus spears on lettuce in a fan shape.

Separate egg whites and yolks. Force egg whites and yolks separately through wire sieve. Sprinkle the whites, then the yolks over asparagus spears in separate bands. Sprinkle with remaining marinade.

YIELD: 4 servings

✳ OL' BOSTON BAKED BEANS ✳

A great do-ahead idea for your next picnic! The beans can be frozen for up to 6 months.

2½ cups dried navy beans, soaked in water overnight (or see Beans, Dried, page 204)	½ cup light molasses
	1 tablespoon soy sauce
	2 teaspoons dry mustard
	½ teaspoon ground black pepper
½ pound unsmoked ham hock, thawed if frozen	1 cup catsup
½ cup chopped onions	1 tablespoon vinegar

Drain beans and place in a medium-size pot with water and bring to a boil over high heat. Lower heat and simmer for 30 minutes.

Drain beans, reserving 1 cup liquid in a separate bowl.

Place beans, ham hock and onions in a 3-quart covered ovenproof casserole or bean pot.

In a small bowl, stir together bean liquid, molasses, soy sauce, mustard, pepper, catsup and vinegar. Pour over beans and mix well.

Bake, covered, at 300°F for 2¾ hours. Remove ham hock. Cool slightly. Remove meat from the bone. Chop meat and stir meat into beans.

Beans can be frozen at this point.

Bake frozen beans straight from the freezer in a 350°F oven for 2½ hours. Or thaw and bake for about 45 minutes until hot and bubbly.

YIELD: 12 to 14 servings, 8 cups

RED BEET AND WALNUT SALAD

Frozen red beets are delicious in this French country salad. The combination of beets and garlic is unusual, but appetizing.

2 cups bite-size pieces frozen cooked red beets, thawed	⅓ cup walnut or olive oil
2 cloves garlic, minced	3 tablespoons red wine vinegar
2 teaspoons finely chopped onions	1 tablespoon honey
½ cup coarsely chopped walnuts	1 small head endive, washed and cut into bite-size pieces

Steam thawed beets for a few minutes to warm.

Toss beets in a glass bowl with garlic, onions, walnuts, oil, vinegar and honey. Marinate at room temperature for at least 2 hours, stirring occasionally. To serve, place endive on serving plate and spoon beet mixture on top.

YIELD: 4 servings

VARIATION: Use green beans instead of walnuts and decrease garlic to 1 clove. Serve over Boston lettuce instead of the endive.

SWEET AND SOUR CABBAGE

Lemon juice sharpens the flavor of this cabbage dish. You can serve it with rice to make a meal, or as a delightful accompaniment to meat or fish.

3 cups coarsely shredded frozen cabbage	2 tablespoons honey
⅓ cup chopped onions	3 tablespoons red wine vinegar
1 apple, cored and diced	1½ teaspoons grated lemon peel
1 very small potato, shredded	juice of 1 lemon

In a large saucepan, cook cabbage and onions in 1 inch of boiling water just until tender, about 6 minutes.

Drain off water and add apple, potato, honey, vinegar, lemon peel and lemon juice. Cover and cook over low heat, stirring occasionally. Heat until apple is tender and sauce is thick, about 10 minutes.

YIELD: 4 servings

NOTE: If the cabbage is frozen in wedges, thaw it slightly, then slice.

VARIATIONS: Substitute 2 cups of chopped kale or 2 cups of green beans for the cabbage. If using green beans, cook the beans as for the cabbage, but use ½ cup thick tomato puree instead of the apple and potato.

BROCCOLI-NOODLE SALAD

2 cups frozen broccoli
  (florets and stems)
2 tablespoons sesame
  seeds, toasted (page 54)
1 teaspoon minced fresh
  gingerroot
1 clove garlic, minced
¼ teaspoon hot red pepper
  flakes
3 tablespoons rice wine
  vinegar
2 tablespoons Chinese
  sesame oil
2 tablespoons vegetable
  stock (page 84) or
  Chicken Stock (page
  73), thawed if frozen
1 tablespoon soy sauce
½ pound Japanese soba
  noodles, wheat or
  buckwheat
1 cup cubed tofu,
  steamed 8 minutes,
  optional

Steam broccoli until just tender. Run under cold water to chill quickly. Cut stems crosswise into thin slices. Trim florets into smaller pieces. Set aside.

Meanwhile, place sesame seeds, gingerroot, garlic, pepper flakes, vinegar, oil, stock and soy sauce in a jar. Cover and shake to combine.

Cook noodles in boiling water until tender, about 8 minutes. Drain and cool under running water. Toss noodles, broccoli, tofu (if desired) and dressing together. Serve at room temperature.

YIELD: 4 to 6 servings

VARIATIONS: The broccoli can be replaced with green beans or cauliflower.

FROZEN CORN-ON-THE-COB

To bring the exciting flavor of freshly picked corn back to the frozen version, serve it with seasoned butter combined with a splash of the cooking water.

4 ears frozen corn, thawed	2 tablespoons butter, at room temperature
¾ cup water	½ teaspoon chili powder

In a covered skillet, cook the corn in the water until tender, 6 to 10 minutes. Check the water to make sure it doesn't evaporate, adding more if necessary, to end up with 2 to 3 tablespoons of corn-flavored water. Remove from heat.

Transfer corn to a hot serving bowl. Add butter and chili powder to remaining water. When butter melts, spoon this seasoned broth over corn and serve at once.

YIELD: 4 servings

VARIATIONS: Substitute for the chili powder 1 teaspoon dried oregano with 1 tablespoon Parmesan cheese, or ¾ teaspoon curry powder or 1 pinch ground black pepper with ⅛ teaspoon ground nutmeg.

OATMEAL-CARROT BARS

To turn this recipe into an extra-special dessert, cut the cookies into squares and top each with a scoop of ice cream or frozen yogurt. The baked bars can be frozen for 3 months.

½ cup molasses	1 cup whole wheat flour
⅓ cup butter	1 teaspoon baking powder
1 egg	½ cup quick-cooking rolled oats
½ teaspoon vanilla extract	¼ cup wheat germ
¾ cup finely shredded carrots	½ cup raisins

In a small mixer bowl, cream together molasses, butter, egg and vanilla with an electric mixer until light and fluffy. Add carrots and mix well.

Thoroughly stir together flour and baking powder. Stir in oats and wheat germ.

Gently stir dry ingredients into creamed mixture; fold in raisins. Spread mixture in a buttered 9 × 9 -inch baking pan. Bake in 350°F oven for 30 to 35 minutes or till golden brown.

Cool in pan on wire rack. Cut into bars or squares and remove from pan.

YIELD: 24 bars

CAULIFLOWER LOAF SOUFFLÉ

¾ cup whole wheat bread crumbs made from slightly stale bread (do not use dry bread crumbs)
2 cups frozen cauliflower florets
4 eggs, separated
1½ cups milk

1 teaspoon fresh thyme leaves
¼ teaspoon grated nutmeg
1 tablespoon finely chopped fresh parsley
dash of cream of tartar
1 cup shredded sharp cheddar cheese

Butter a 9 × 5-inch loaf pan. Sprinkle 2 tablespoons of the bread crumbs in the pan and shake to distribute the crumbs evenly.

Steam the cauliflower until tender, about 10 minutes. Cool and cut into thin slices.

Meanwhile, beat egg yolks until thickened. Stir in remaining bread crumbs, milk, thyme, nutmeg and parsley. Add cauliflower.

Beat the egg whites with the cream of tartar until soft peaks form.

Stir half of the egg whites into the cauliflower mixture. Fold in the remaining egg whites and the cheese. Pour into the prepared pan.

Bake in a preheated 400°F oven for 20 minutes. Reduce heat to 350°F and continue baking until top is browned, 20 to 25 minutes more. Serve immediately.

YIELD: 4 to 6 servings

VARIATIONS: The cauliflower can be replaced with either broccoli or carrots.

BABA GHANNOUJ

This is a Middle Eastern eggplant puree flavored with tahini, lemon juice and garlic. This version should not be refrozen.

2 cloves garlic	2 tablespoons olive oil
1 cup frozen roasted eggplant pulp (page 208), thawed and drained	1 teaspoon soy sauce
	¼ cup fresh parsley leaves
	olive oil
	chopped parsley
½ cup tahini	pita bread (page 310)
2-3 tablespoons lemon juice	

Food Processor Instructions
Using metal blade, with machine on, drop garlic into machine and process until finely chopped. Add eggplant and process until smooth. Add tahini, 2 tablespoons of the lemon juice, oil and soy sauce. Process until very smooth. Add parsley and process until slightly chopped. Taste and add extra lemon juice if needed.

Blender Instructions
Place garlic, 2 tablespoons lemon juice and oil in blender. Process until garlic is finely chopped. Add eggplant, tahini and soy sauce. Blend until smooth, stopping occasionally to scape the sides of the container. Add parsley and blend until slightly chopped. Taste and add extra lemon juice if needed.

Let mixture stand in the refrigerator for at least 2 hours before serving. To serve, spread in a shallow bowl and drizzle with olive oil. Sprinkle with chopped parsley and surround with triangles of pita bread.

YIELD: 1½ cups

EGGPLANT AND BASIL GRATIN

Frozen eggplant slices are delicious in this casserole. If you do not have fresh or butter- or oil-preserved basil, substitute parsley mixed with 2 teaspoons dried basil. If made with frozen eggplant, this dish should not be refrozen. If made with fresh eggplant, you can freeze it for 6 months.

olive oil
16 ½-inch-thick peeled
      eggplant slices, thawed
      if frozen
¾ cup peeled, seeded and
      drained tomatoes,
      thawed if frozen
½ cup ricotta cheese
½ cup shredded mozzarella
      cheese
1 egg

2 cloves garlic, minced
      pinch of cayenne pepper
½ cup grated Parmesan
      cheese
½ cup yogurt
½ cup coarsely chopped
      fresh basil, or ¼ cup
      basil preserved in olive
      oil or butter, at room
      temperature
¼ cup bread crumbs

Brush a baking sheet with oil. Place the eggplant slices on the tray and brush the tops with oil. Broil until golden brown on top, then turn the slices over and continue broiling until tender. (Add more oil if necessary to keep eggplant from sticking.) Set aside.

In a skillet, simmer tomatoes gently until very thick. Set aside.

In a bowl, stir together ricotta and mozzarella cheeses, egg, garlic, cayenne, ¼ cup of the Parmesan cheese and the yogurt.

To assemble, place half of the eggplant slices in the bottom of an 8 × 8-inch baking dish. Sprinkle fresh basil over the eggplant or spread the preserved basil over eggplant. Spoon and smooth the tomato pulp over the basil. Top with the remaining eggplant.

Pour the cheese and egg mixture over the top and gently tap the dish on a hard surface to settle the liquid down and into the eggplant.

Mix the bread crumbs and remaining Parmesan cheese and sprinkle over the surface. Cover and freeze now, or bake at 350°F until lightly browned, about 35 minutes.

If baking a frozen casserole, cover loosely and bake for 50 minutes. Uncover and bake until browned, another 15 to 20 minutes.

YIELD: 4 to 6 servings

EGGPLANT-CHEESE BAKE

A cross between lasagna and eggplant parmesan, this oven-ready recipe is a good way to freeze eggplant for up to 4 months. It is also good when prepared with frozen eggplant, but then it should not be refrozen.

1 medium-large eggplant, peeled and cut into ½-inch slices, or 14 slices of frozen eggplant	2 teaspoons fresh thyme leaves, or 1 teaspoon dried thyme
¾ cup water	¼ cup tomato paste, optional
⅓ cup finely chopped onions	2 cups ricotta cheese
3 cloves garlic, minced	2 cups shredded Monterey Jack cheese
1 tablespoon olive oil	½ cup whole wheat bread crumbs or wheat germ
2 cups tomato puree, thawed if frozen	

Place eggplant in a large skillet with water, cover tightly and simmer over low heat for 6 to 8 minutes or until the eggplant is softened. Drain and reserve the eggplant.

Cook onions and garlic in oil over medium heat until limp. Add tomato puree and thyme and simmer for 5 minutes. If puree is thin, thicken with tomato paste.

Arrange half of the eggplant in the bottom of an oiled 8½ × 4½-inch loaf pan or a 1-quart baking dish. Top with half of the ricotta, then half of the tomato sauce. Repeat the layering.

Sprinkle with Monterey Jack cheese and bread crumbs. Cover and freeze at this point, or if preparing right away, bake at 375°F for 40 minutes or until bubbly and browned on top.

To prepare frozen Eggplant Cheese Bake, bake at 350°F, covered, for 45 minutes. Uncover and bake until brown and bubbly, about 15 minutes more.

YIELD: 4 to 6 servings

VARIATIONS: Almost any vegetable is good when paired with ricotta cheese and tomatoes. Instead of the eggplant, try 4 cups of cooked, coarsely chopped broccoli; 3 cups of chunked or mashed, tender-cooked winter squash; or 4 cups of zucchini slices steamed and then drained.

GREENS WITH GARLIC AND OIL

We think this straightforward and aromatic preparation is the best way to bring out the virtues of frozen spinach, kale, beet greens, turnip greens or almost any other leafy green vegetable. These greens are especially good served with grilled or baked fish, chicken, beef or lamb.

2 cups frozen greens, thawed and coarsely chopped	2 tablespoons olive oil
	1 clove garlic, minced
	1 tablespoon lemon juice

Cook greens in their own liquid until tender. Drain if very wet.

Warm oil gently in a skillet. Add garlic and cook about 30 seconds. Add the greens and lemon juice and stir in a few seconds until heated through.

YIELD: 4 servings

VARIATION: Substitute 2 cups cooked broccoli florets for the greens and add ½ teaspon dried thyme to oil with the garlic.

PEAS AND MINT

To perk up the flavor of frozen peas, steam them over mint herb tea.

1½ cups water	2 cups frozen peas
2 bags herbal mint tea, or 4 teaspoons dried mint leaves	1 tablespoon butter

Put water in a saucepan. Bring to a very slow boil and add tea bags. Place a steamer rack in the pan, add frozen peas, cover and cook until peas are bright green and tender—about 10 minutes. Regulate the heat to avoid a full boil which would damage the flavor of the mint.

Transfer peas to a warm bowl and toss with butter and a spoonful of the tea.

YIELD: 4 servings

CHEESE-STUFFED PEPPERS,
PLOVDIV STYLE

This Bulgarian dish is very similar to Mexican chilis rellenos. Sprinkling the peppers with ground spices derives from the Turkish influence in this cuisine. This dish can be prepared and frozen for up to 4 months, or the filling can be used with frozen peppers.

24 hot Hungarian wax peppers, or 10 sweet Italian frying peppers, roasted and peeled	¼ cup water
	⅔ cup whole wheat pastry flour
	sunflower oil for frying
3 cups diced Bulgarian feta cheese, about 16 ounces	1 tablespoon ground coriander seeds
4 eggs, separated	2 teaspoons ground cumin seeds

Cut a slit from end to end in each pepper. Remove seeds.

Mix cheese with two of the egg yolks and set aside. Mix remaining egg yolks well with water. Stir in ¼ cup of the flour to form a smooth pancakelike batter.

Stuff peppers with cheese mixture, pressing the peppers back into their original shape. Dredge peppers lightly in remaining flour, then shake gently to remove excess flour.

Beat egg whites until soft peaks form. Fold egg whites into batter until evenly mixed.

Heat a ½-inch-deep layer of oil in a large skillet. A spoonful of batter dropped in the oil should sizzle and brown slowly when the oil is the right temperature. The oil should be hot enough to cook the batter before the filling gets hot.

Using tongs, dip peppers in batter and ease into skillet. Using tongs and a fork, turn peppers to brown all sides. Drain briefly on paper towels or brown paper bags. Cool and freeze, or if serving immediately, bake in 350°F oven until very hot throughout, about 15 minutes.

Meanwhile, combine coriander and cumin in a small bowl.

Pass the spice mixture at the table to sprinkle over the peppers.

To prepare from the freezer, place directly on baking sheet and bake, uncovered, in a 350°F oven, until very hot, about 35 minutes.

YIELD: 8 to 12 servings

NOTE: Bulgarian feta cheese is available in small cans or by the piece. Soak cheese in fresh water for one day before you use it. If you prefer, substitute hard "Salata" ricotta or Greek feta cheese.

VARIATIONS: For a Mexican flavor, use roasted Anaheim peppers, and stuff them with strips of Monterey Jack cheese.

Corn or ground meat can also be used in the stuffing. Dip in batter and cook as directed.

SAUTÉED PEPPERS AND ONIONS

Use a colorful combination of bell and frying peppers in both red and green to make this elegant side dish. You can vary the shapes in which you cut the peppers, as well as the varieties of pepper you use. Freeze for 6 months.

2 tablespoons olive or
      vegetable oil
2 cloves garlic, minced
4 cups pepper strips
¾ cup sliced onions
½ pound mushrooms, sliced
2 teaspoons chopped fresh
      basil, or 1 teaspoon
      dried basil

1½ teaspoons fresh tarragon,
      or 1 teaspoon dried
      tarragon
2 teaspoons soy sauce

In a large skillet, heat the oil and add garlic, peppers and onions. Sauté over medium heat for about 3 minutes. Add mushrooms, basil and tarragon and sauté about 4 more minutes. Add soy sauce and cook about 3 minutes for a firm texture. Serve as a side dish or add to soup.

To serve this dish as a sauce, continue cooking until peppers are very tender, about 10 minutes.

YIELD: 3 cups

CRESPELLE STUFFED WITH SQUASH

Crespelle are thin, egg-rich Italian pancakes. The combination of tomato sauce and sweet squash filling is both beautiful and flavorful. The crespelle and the sauce are versatile and you may want to make larger batches for the freezer. Unfilled crespelle can be frozen for 6 months. Filled crespelle can be frozen for 2 months.

Crespelle

3 tablespoons whole wheat pastry flour
3 eggs, lightly beaten
3 tablespoons milk
1 teaspoon grated orange peel

1 tablespoon safflower oil butter for cooking crespelle

Sauce

⅓ cup butter
4 cups coarsely chopped, peeled and drained plum tomatoes, thawed if frozen

12 fresh basil leaves, coarsely chopped, or 1 teaspoon dried basil

Filling

1½ cups ricotta cheese
1 cup shredded mozzarella cheese
1 egg
⅛ teaspoon freshly ground black pepper

¼ teaspoon freshly grated nutmeg
1½ cups yellow squash puree, thawed if frozen (a dry-fleshed squash is best)

To make the crespelle, combine flour and eggs with a wire whisk. (A few lumps are all right.) Slowly stir in milk, orange peel and oil. Cover and refrigerate for at least 1 hour before cooking. Meanwhile, begin the sauce, which can be stirred while cooking the crespelle.

To make the sauce, melt butter in a medium-size saucepan over medium heat. When butter foams, add tomatoes and crush with a wooden spoon. Add basil. Simmer, uncovered, to form a thick sauce. If tomatoes were well drained this should take 15 to 20 minutes. Puree in a blender if a smoother sauce is desired. Keep hot.

When ready to cook the crespelle, heat butter over medium to low heat in an 8-inch skillet with sloping sides. When bubbly, tilt pan slightly and swirl 2 tablespoons batter in a thin layer over the bottom of the pan. When lightly browned on the bottom, flip over to cook other side. When lightly browned, remove from the pan. Repeat process to use remaining batter, replacing butter when needed. Cool crespelle in a single layer, best side down. This batter should make 10 to 12 crespelle.

To make the filling, beat the ricotta and mozzarella cheeses with the egg, pepper and nutmeg in a bowl. Fold in the squash.

Place 3 tablespoons of filling in a strip down the middle of each crespelle. Fold sides over the filling.

Freeze now or place the filled crespelle in a 9 × 13-inch buttered ovenproof casserole. Bake in a 375°F oven until filling is very hot, 25 to 35 minutes. Let stand for 3 minutes before serving. Top with hot tomato sauce.

To heat filled and frozen crespelle, bake loosely covered in a 350°F oven until very hot, about 45 minutes, then top with hot tomato sauce.

YIELD: 10 to 12 crespelle

VARIATIONS: Substitute 1 tablespoon basil butter (page 100) for the fresh or dried basil in the sauce and decrease the butter by 1 tablespoon.

Similar ingredients can be assembled into an easy casserole: layer pureed squash with cooked noodles or rice, cheese and tomatoes.

SORREL PUREE

Use the following proportions with any amount of sorrel. Its volume reduces greatly—3 cups raw leaves yields ⅓ cup cooked—but since the flavor is intense it goes a long way. You can freeze this puree for 9 months.

1 clove garlic, minced
2 shallots, finely chopped
2 tablespoons butter

3 cups lightly packed
    sorrel leaves (stems
    removed)

In a saucepan over medium heat, cook the garlic and shallots in butter for 2 minutes. Add sorrel leaves and cook, stirring occasionally, until sorrel is very soft—it actually falls apart. Cool and freeze.

YIELD: ⅓ cup

MEDITERRANEAN SPINACH CASSEROLE

In the south of France and along the Spanish coast near Barcelona, wonderful spinach casseroles using raisins or figs and pine nuts are a staple side dish. This one can be frozen for 6 months.

½ cup chopped onions
2 cloves garlic, minced
2 tablespoons olive oil
2 cups coarsely chopped spinach, packed
2 eggs, well beaten
½ cup raisins

2 teaspoons lemon juice
½ cup pine nuts
¼ cup shredded mozzarella cheese
1 tablespoon grated Parmesan cheese

In a large skillet, sauté the onions and garlic in oil until onions are soft and translucent, about 5 minutes. Add spinach. Cook to evaporate extra liquid, about 8 minutes.

Allow spinach to cool slightly, then add eggs, raisins, lemon juice and nuts. Mix well and spread into a lightly buttered 1-quart ovenproof casserole.

Freeze now, or sprinkle the top with cheeses, cover and bake 20 minutes at 350°F. Remove cover and bake 5 to 8 minutes more.

To bake a frozen casserole, place it covered in a 350°F oven and bake until hot, about 45 minutes. Uncover, sprinkle the top with cheeses, and bake for 5 to 8 minutes more.

YIELD: 6 to 8 servings

NOTE: To use frozen spinach, thaw and drain the spinach thoroughly and add to recipe as specified above.

VARIATION: This casserole can also be prepared with other leafy greens such as chard, Chinese cabbage or kale.

SQUASH AND EGG CASSEROLE

Serve this Transylvanian casserole as a light main course or a vegetable side dish. This recipe assumes that the squash was cooked until almost tender before freezing.

3 cups frozen yellow
squash slices, about 1
pound, thawed enough
to separate
½ cup diced onions
2 tablespoons butter
4 eggs
1 tablespoon whole wheat
pastry flour

½ cup sour cream
⅛ teaspoon ground
nutmeg
2 teaspoons chopped fresh
dillweed
¼ cup grated Parmesan
cheese

Sauté squash and onions in butter over medium heat, turning to brown both sides evenly. Squash should be tender; if it's not, add a spoonful of water, cover and steam until tender. Transfer to a buttered 8 × 8-inch baking dish.

Meanwhile, beat 1 egg with flour until smooth. Mix in remaining eggs, sour cream, nutmeg, dillweed and Parmesan cheese.

Pour egg mixture over squash and bake in 350°F oven for 15 to 20 minutes, until eggs are just set and are slightly puffy. Serve while hot.

YIELD: 4 to 6 servings

NOTE: To substitute fresh squash, peel and seed a 1½-pound squash. Cut into slices and brown in butter. Add onions and transfer to casserole. Bake in 350°F oven until squash is tender, about 15 minutes. Then proceed as directed above with the egg mixture.

HUBBARD SQUASH WITH CRANBERRIES, ORANGE AND WALNUTS

The cranberries sit like ruby jewels in their golden squash nest. If made with fresh squash, you can freeze this for 6 months. It won't freeze as long if made with frozen squash.

3 cups mashed or pureed
fine-textured winter
squash (Hubbard or
buttercup), thawed
if frozen
1 egg, beaten
¼ cup melted butter

¾ cup cranberries
¼ cup frozen unsweetened
orange juice concen-
trate, thawed
2 tablespoons honey
pinch of ground nutmeg
½ cup chopped walnuts

If the squash is very cold, warm it gently in a saucepan. Stir egg and 2 tablespoons of butter into the squash.

In a separate bowl, stir together the cranberries, orange juice concentrate, honey and nutmeg. Stir ¼ cup of this mixture into the squash. Spoon squash into a buttered 1-quart ovenproof casserole and top with remaining cranberry mixture. Freeze now or bake at 400°F until sizzling hot, about 25 minutes. Sprinkle with walnuts and serve.

To prepare frozen casserole, bake at 350°F, loosely covered, until very hot, about 1 hour.

YIELD: 4 to 6 servings

FROZEN PUMPKIN MOUSSE

This festive dessert, frozen in a decorative mold, is a fitting finale to a holiday dinner. It can be frozen for 2 to 4 months.

¾ cup pumpkin or squash
puree, thawed if frozen
¼ cup wheat germ
1 teaspoon ground
cinnamon
¼ teaspoon ground ginger

¼ teaspoon ground mace
¼ cup honey
¼ cup molasses
3 eggs, separated
½ cup heavy cream,
whipped

In a heavy saucepan, mix together pumpkin puree, wheat germ, cinnamon, ginger, mace, honey and molasses. Add egg yolks and stir constantly over low heat until thickened, 10 to 15 minutes. Remove from stove and let cool to room temperature.

Beat egg whites until stiff and gently fold into the pumpkin mixture with a rubber spatula.

Then fold in whipped cream and spoon into a 1-quart decorative mold. Place mold in freezer until firm, about 4 hours. To unmold, dip briefly in hot water and invert onto a chilled serving plate. Uneaten portion can be kept frozen if covered, but texture will become slightly grainy after 2 days.

YIELD: 6 to 8 servings

ZUCCHINI PUREE

This puree is used as a thickening base for soups and stews—an excellent use for zucchini. The mild flavor makes it an excellent base for spaghetti sauce as well. You can freeze the puree for 9 months.

1 cup water	3 onions, sliced
10 fresh zucchini, cubed	4 red bell peppers, seeded and sliced

Combine water, zucchini, onions and peppers in a large pot. Cover and simmer until all vegetables are tender. Place vegetables and liquid 1 cup at a time into the blender and puree. Freeze in 2-cup portions.

Before using, thaw at room temperature or by cooking slowly in a covered saucepan.

YIELD: about 14 cups

GREEN TOMATO "MINCEMEAT"

Green tomatoes have long been used as a pie filling, either "as is" or prepared as a substitute for conventional mincemeat. This mincement can be frozen for 9 months.

2 pounds fresh green tomatoes	¾ cup light molasses
4 tart baking apples, cored	1 cup raisins
4 pears, cored	2 teaspoons ground cinnamon
1 lemon	½ teaspoon ground nutmeg
¼ cup vegetable oil	¼ teaspoon ground ginger
½ cup white grape juice	½ teaspoon ground allspice
2 tablespoons cider vinegar	

Chop tomatoes, apples and pears into ¼-inch pieces, catching any juice that runs out.

Peel and juice the lemon. Chop the lemon peel into small pieces.

In a large pot, warm oil over medium heat and add tomatoes, apples, pears, lemon peel and juice, grape juice, vinegar and molasses. Cook, uncovered, stirring occasionally until the mixture has a very thick, soupy consistency, 30 to 40 minutes. Add raisins, cinnamon, nutmeg, ginger and allspice. Taste and, if necessary, add more molasses or vinegar.

Stirring frequently over low heat, reduce until very thick, about 1 hour. When finished, the "mincemeat" should hold its shape when stirred with a spoon. Cool and freeze in 4-cup portions.

To bake a mincemeat pie, thaw filling and pour into an unbaked 9-inch whole wheat pie shell (page 329). Cover with a top crust or crumbs. Bake in a 375°F oven for 45 to 55 minutes, until crust browns lightly.

YIELD: about 8 cups, enough for two 9-inch pies

RUSSIAN SALAD

This is a real catch-all salad. The proportions and ingredients can be varied to suit the vegetables at hand.

2 medium-size red-skinned potatoes
1 cup frozen diced or thinly sliced carrots
1 cup frozen peas
1 cup frozen diced red beets, thawed
¼ cup finely chopped celery
¼ cup finely chopped scallions, including green part

2 teaspoons chopped fresh dillweed, or
¾ teaspoon dried dillweed
1 tablespoon chopped fresh parsley
½ cup yogurt
¼ cup mayonnaise ground black pepper, to taste

Steam the potatoes over boiling water, cool slightly, then dice and place in a large bowl. Steam carrots until barely tender and add to potatoes. Steam peas until tender but still bright green. Rinse under cold water to help preserve the color. Drain the peas thoroughly and add to potatoes along with beets, celery, scallions, dillweed and parsley.

In a small bowl, stir together yogurt, mayonnaise and pepper. Pour over salad and toss gently until well mixed. Refrigerate until cold. Serve on a bed of lettuce or in cabbage-leaf cups.

YIELD: 6 servings

VARIATIONS: Add 1 teaspoon curry powder to yogurt-mayonnaise mixture to replace the dillweed.

For eggs casino, add 3 tablespoons pickle relish and 2 teaspoons prepared mustard and an additional ½ cup mayonnaise to the yogurt-mayonnaise mixture. Cut four hard-cooked eggs in half. Mash the yolks with ¼ cup of the yogurt-mayonnaise mixture and use to fill four of the egg white halves. Press the remaining egg white halves over the stuffing so that the eggs look whole. Place the eggs on top of Russian Salad. Spoon remaining mayonnaise mixture over eggs and serve.

RATATOUILLE

Capture the summery herbal aroma of this French vegetable stew to enjoy year-round. The vegetables in the best ratatouilles are cooked separately and assembled in layers for final cooking. This ratatouille will freeze for 9 months.

¼ cup olive and vegetable oil, mixed in equal amounts

2 large eggplants, cut into ¾-inch cubes, about 6 cups

2 small zucchini, halved lengthwise, then sliced crosswise

2 teaspoons chopped fresh rosemary, or 1 teaspoon dried, crumbled rosemary

4–6 cloves garlic, minced

1 Spanish onion, sliced and cut to 1-inch lengths, about 2½ cups

4 bell peppers, red and green, chopped, about 5 cups

2 tablespoons finely chopped fresh hot chili peppers, optional

2 teaspoons fresh thyme leaves, or 1 teaspoon dried thyme

4 cups chopped peeled tomatoes, with juice

1 teaspoon dried oregano

¼ cup lemon juice

¼ cup chopped fresh basil, or 2 teaspoons dried basil

¼ cup, per serving, shredded Gruyère cheese, optional

Heat a thin layer of the oil in a stockpot. Over medium-high heat, sauté eggplant until lightly browned, adding more oil if necessary. Add zucchini, rosemary and half of the garlic. Cover loosely and reduce heat. Simmer until eggplant and zucchini are tender, about 10 minutes. Divide mixture into three 8 × 8-inch baking dishes, about a 1-inch-thick layer in each dish.

In the same pot, heat remaining oil over medium-high heat and sauté onions, bell peppers and chili peppers for 3 minutes. Add thyme and remaining garlic. Cover loosely and reduce heat. Simmer until peppers are tender, about 8 minutes. Spread this mixture on top of eggplant-zucchini layer.

In the same pot, bring the tomatoes and oregano to a boil. Remove from heat, stir in lemon juice and basil. Pour evenly over onion-pepper layer. Shake pans gently to distribute ingredients. Sprinkle with cheese if serving immediately or cool, cover tightly and freeze.

Bake frozen ratatouille, covered, at 375°F for 30 minutes. Uncover and continue baking for 40 minutes more. If desired, top with cheese and bake until cheese just melts.

If ratatouille was frozen in freezer containers, thaw, transfer to an ovenproof casserole or saucepan, then bake or heat on top of stove.

YIELD: 12 to 15 servings, about 4 quarts

SOUTHERN VEGETABLE STEW

Freeze this stew for 1 year if the vegetables are fresh, 6 months if made with frozen vegetables.

¾ cup chopped onions	¼-½ teaspoon red pepper
5 cloves garlic, minced	flakes
¼ cup olive oil	4 cups lima beans
6 cups coarsely chopped	4 cups corn
tomatoes, thawed	4 cups sliced okra
if frozen	
½ teaspoon ground nutmeg	
1 tablespoon chopped	
fresh basil, or 2 tea-	
spoons dried basil	

In a very large pot, cook onions and garlic in oil until limp. Add tomatoes, nutmeg, basil and pepper flakes. Simmer for 15 minutes. Add beans, corn and okra and simmer until almost tender (or completely tender if serving right away), about 20 minutes. Freeze in serving-size portions.

To cook, thaw almost completely, then heat in a saucepan over low heat, cooking until vegetables are very tender and stew is heated through.

YIELD: 8 to 12 servings, 6 to 7 pints

NOTE: This may be served as a meatless main course or as an accompaniment to simple meat dishes. It is especially good with rice or spoon bread.

VEGETABLE TERRINE

Frozen vegetables can be used in very fancy ways, as this attractive, multilayer terrine proves.

2 cups cooked brown rice, thawed if frozen, at room temperature
1 cup shredded cheddar, Gruyère or Jarlsberg cheese
6 tablespoons grated Parmesan cheese
¼ teaspoon ground black pepper
2 eggs, beaten
2 scallions, finely chopped
1 onion, finely chopped
¼ cup butter

1 clove garlic, minced
2 cups frozen chopped spinach or chard, thawed and drained well
1 teaspoon curry powder
¼ cup fine dry bread crumbs
1 cup frozen mashed yellow winter squash, thawed
½ cup sour cream
½ cup chopped walnuts, optional

Prepare an 8½ × 4½-inch loaf pan by cutting a piece of wax paper to fit the bottom. Butter the bottom of the pan, place sheet of wax paper in the pan and set aside.

In a medium-size bowl, mix rice with ½ cup cheddar cheese, 3 tablespoons Parmesan cheese, ⅛ teaspoon pepper, 2 tablespoons beaten egg and the scallions. Press half of this mixture into the bottom of the loaf pan.

In a large skillet, sauté onions in 2 tablespoons butter until tender, about 3 minutes. Add garlic and spinach. Sauté until nearly all excess moisture evaporates, about 5 minutes. Transfer onion, garlic and spinach to a bowl and set in a larger bowl filled with ice water. When spinach mixture is cool, stir in ½ teaspoon curry powder, 2 tablespoons beaten egg, 2 tablespoons bread crumbs, ¼ cup cheddar cheese and 3 tablespoons Parmesan cheese. Spread spinach mixture over rice layer.

Mix squash with 2 tablespoons beaten egg, 2 tablespoons bread crumbs, dash of black pepper, ½ teaspoon curry powder and ¼ cup cheddar. Spread over spinach layer.

Top squash layer with remaining rice mixture. Dot with butter. Cover loosely with aluminum foil and bake in a 350°F oven for 1 to 1¼ hours or until firm and edges begin to brown. (Knife inserted in center will come out clean.) Chill completely.

Before serving, run knife around edges of terrine. Place pan in bowl of hot water for 30 seconds. Unmold and remove wax paper. Cut into slices and serve with sour cream.

Sprinkle with chopped walnuts, if desired.

YIELD: 8 to 16 servings

VARIATIONS: For a less spicy terrine, substitute ½ teaspoon nutmeg or thyme for the curry powder.

Use 2 cups finely chopped frozen broccoli, thawed and drained well, in place of spinach.

Cooked corn or peas can be added to the rice layer for extra color.

GINGER-VEGETABLE STIR-FRY

While unusual, stir-frying is possible with frozen vegetables if you thaw them only long enough to separate the pieces. The texture of some vegetables will be less crisp, but cooking is still fast, with delicious possibilities.

3 tablespoons vegetable oil	1 cup coarsely chopped
½ cup chopped onions	frozen greens, such as
2 cloves garlic, minced	spinach or beet greens,
12 mushrooms, sliced	thawed completely
4 cups chopped frozen	1 tablespoon minced fresh
vegetables, such as	gingerroot
beans, red sweet	2 teaspoons soy sauce
peppers or broccoli,	pinch of cayenne pepper
slightly thawed	

Heat the oil for a few seconds in a wok or large, heavy skillet. Add onions and stir-fry over high heat for about 1 minute. Stir in garlic, mushrooms and vegetables. Toss while cooking for 2 minutes. Add greens and gingerroot. Lower heat, cover and simmer until the vegetables are just tender. Add soy sauce and cayenne. Stir and serve immediately over hot brown rice.

YIELD: 4 to 6 servings

VEGETABLE PÂTÉ

This light and fresh-tasting pâté is especially good served with mayonnaise or a yogurt-based sauce. It's delicious made with fresh or frozen vegetables.

½ cup chopped onions	½ teaspoon fresh rosemary
2 cloves garlic, minced	pinch of red pepper
2 tablespoons butter	flakes
½ pound mushrooms, finely chopped	2 tablespoons lemon juice
	3 eggs
2 cups frozen carrot slices, partially thawed	¼ cup shredded Swiss or Gruyère cheese
2 tablespoons water	1 cup frozen, thawed, drained and packed spinach
1 teaspoon fresh thyme leaves	
1 teaspoon chopped fresh basil	1 bay leaf

Sauté onions and garlic in butter in a medium-size saucepan until soft and translucent. Add mushrooms and carrots and cook over medium-high heat about 5 minutes. Add water to pan, cover and simmer about 5 more minutes or until carrots are tender. Stir in thyme, basil, rosemary, pepper flakes and lemon juice and transfer mixture to a food processor or blender. Process for a few seconds. Then, with motor running, add the eggs one by one, then the cheese. When completely mixed, remove work bowl from machine and stir in spinach by hand.

Thickly butter a 9 × 5-inch loaf pan. Add the pâté mixture to the loaf pan. Place bay leaf on top, cover with foil and bake at 350°F for 30 minutes. Uncover and bake 10 minutes more or until firm.

Chill several hours or overnight. Run a knife around edges of the pan and unmold onto a plate. Cut in slices and serve.

YIELD: 6 to 8 servings

VARIATIONS: You can use cooked, frozen green beans or broccoli instead of the spinach.

GARDENER'S PANCAKES

This Hungarian recipe features frozen vegetables layered with crespelle, cheese and sour cream. It is baked and served in large wedges like cake. The vegetable selection is unlimited. The vegetables should be cooked but can be hot or cold when assembling the "cake" (the baking time is slightly shorter if vegetables are still hot).

2 cups chopped frying
   peppers
2 tablespoons butter
½ cup finely chopped
   onions
8 crespelle (page 228),
   thawed if frozen
1 cup cooked 1-inch pieces
   of green beans
1 cup sour cream,
   approximately

2 cups shredded Jarlsberg
   cheese
1 cup frozen cooked diced
   carrots or diced winter
   squash, thawed
1 cup frozen cooked finely
   chopped broccoli,
   thawed
  Hungarian paprika

Sauté the peppers in butter until tender. Stir in the onions and set aside.

Butter two pie pans and place a crespelle in each. Divide the beans between the crespelle and sprinkle 2 tablespoons of the pepper mixture over the beans.

Stir sour cream until it is smooth, then spread some of it thinly over the beans and peppers. Sprinkle with some cheese and top with another crespelle.

Divide the carrots over the crespelle and follow with pepper mixture, sour cream, cheese and another crespelle. Repeat layers using broccoli.

Sprinkle top crespelle with cheese and paprika. Bake until bubbly hot in a 375°F oven, about 30 minutes. Let stand 5 minutes before cutting. Cut into large wedges with a sharp knife.

YIELD: 4 to 6 servings

VARIATIONS: Add a bit of hot pepper to the frying peppers for a zippy taste.

Substitute lima beans for the green beans.

FROZEN GARDEN MEDLEY

The following recipe is made of vegetables that come into season at the same time. Leaving this simple stew unseasoned when you freeze it will make it more versatile later. Add garlic, oregano and thyme for a Mediterranean flavor, or use chili powder for a hearty Southwestern aspect.

The vegetables are "blanched" by cooking them in the tomato liquid and can be kept somewhat crisp by chilling filled freezer containers in a sink filled with ice water. Freeze for up to a year.

5 quarts fresh tomatoes, peeled and quartered, about 22 medium-size tomatoes	1 cup diced zucchini
	1 onion, diced
	1 cup okra pods, thinly sliced
1 cup diced yellow summer squash	1 cup cauliflower florets

Bring tomatoes to a boil in a large covered pot, stirring occasionally. Add yellow squash, zucchini, onions, okra and cauliflower.

Boil gently for 3 minutes. Transfer to freezer containers and chill, then freeze.

To serve, bring slowly to a boil, add any seasonings desired, and cook until vegetables are tender-crisp, about 8 minutes. Serve as is or over broken crackers, rice or noodles. Top with shredded cheese if desired.

YIELD: 16 servings, 4 quarts

VARIATION: Substitute 1 cup green beans cut in 1-inch lengths for the cauliflower.

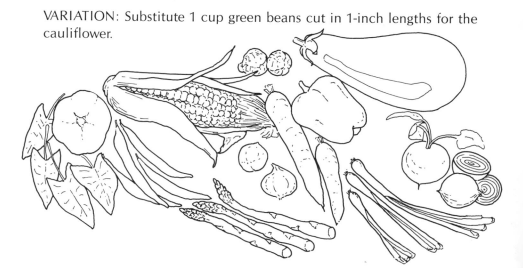

VEGETABLES MÉLANGE

This recipe is an excellent way to use frozen vegetables. Basically a stir-fry, the blanched vegetables need little additional cooking. This dish can be frozen for 8 months.

2 tablespoons olive oil
⅔ cup ¼-inch sliced carrots
1 cup broccoli florets and crosscut stems
1 cup sliced zucchini, thawed if frozen
½ cup chopped green peppers
1½ cups mushrooms, sliced
¼ cup sliced scallions, including green tops
2 tablespoons finely chopped fresh parsley

1 tablespoon finely chopped fresh basil, or 1 teaspoon dried basil
1 teaspoon dried oregano
½ teaspoon fresh thyme leaves, or ¼ teaspoon dried thyme
cayenne pepper, to taste
1½ cups crushed tomatoes, thawed if frozen
¼ cup grated Parmesan cheese

Heat oil in a skillet over medium heat, about 3 minutes. Increase heat to medium high, add carrots and broccoli and cook for 5 minutes, stirring occasionally. Add zucchini, peppers, mushrooms and scallions. Cook for 5 minutes more, stirring occasionally. Stir in parsley, basil, oregano, thyme, cayenne and tomatoes and cook until heated through, about 5 minutes. Freeze now in rigid containers or serve immediately, sprinkled with Parmesan cheese.

To use frozen dish, thaw in the refrigerator before heating. Sprinkle with Parmesan cheese.

Serve over hot brown rice, noodles, potatoes or polenta.

YIELD: 4 servings, 4 cups

VEGETABLES BAKED
IN CHEESE SAUCE

This casserole can be frozen for 4 months if you make the sauce with Mochiko rice flour.

3 large leeks, trimmed, cut in half and then into 1-inch lengths, or 2 cups cooked leeks, thawed if frozen

3 cups small cauliflower florets

2 tablespoons butter

3 tablespoons whole wheat pastry flour or, if freezing, 2 tablespoons Mochiko rice flour

1¾ cups milk

¾ cup shredded sharp cheddar cheese

freshly ground black pepper, to taste

3 tablespoons grated Parmesan cheese

ground nutmeg

paprika

Steam leeks and cauliflower separately until almost tender. Arrange leeks and cauliflower in a wide, buttered 1½- or 2-quart ovenproof casserole.

Melt butter in small saucepan over medium-low heat. Add flour and cook for about 2 minutes, stirring constantly.

Add the milk, a little at a time, while stirring with a wire whisk. When slightly thickened, remove from heat and stir in the cheddar cheese and black pepper.

Pour sauce over leeks and cauliflower. Sprinkle with Parmesan cheese, nutmeg and paprika. Freeze at this point or bake at 350°F until lightly browned, 20 to 30 minutes. To prepare frozen casserole, bake an additional 30 to 40 minutes.

YIELD: 4 to 6 servings

VARIATIONS: Replace leeks with pearl onions and replace cauliflower with asparagus, broccoli, chard, kale or parsnips, or a combination of them.

FROZEN VEGETABLE STIR-FRY

Blanched vegetables require less cooking time than fresh, so this stir-fry is especially quick to prepare. Toasted sunflower seeds add extra crunch. You won't get very crisp vegetables, but the texture is crunchy and the cooking technique fast.

3 tablespoons vegetable oil
1 cup frozen ¼-inch-sliced carrots
2 cups frozen chopped broccoli, partially thawed
½ cup chopped red bell peppers
½ cup sliced mushrooms
1 cup frozen corn, thawed and drained
1 cup frozen thinly sliced zucchini, thawed and well drained
2 scallions, chopped

2 tablespoons finely chopped fresh parsley
1 tablespoon finely chopped fresh basil, or 1 teaspoon dried basil
1 teaspoon dried oregano
½ teaspoon fresh thyme leaves, or ¼ teaspoon dried thyme
cayenne pepper, to taste
1 teaspoon soy sauce, optional
2 tablespoons sunflower or sesame seeds, toasted, optional (page 54)

Heat oil in a large skillet over medium-high heat. Add carrots. Cook about 3 minutes, stirring constantly. Carefully add broccoli, peppers and mushrooms (thawed vegetables will splatter). Continue stirring and cooking until vegetables are almost tender and mushrooms are light brown, 3 to 5 minutes.

Add corn, zucchini, scallions, parsley, basil, oregano, thyme and cayenne. Stir and sauté for 2 minutes longer. Add soy sauce if additional seasoning is desired. Sprinkle with sunflower or sesame seeds before serving.

YIELD: 3 to 4 servings

VARIATION: Substitute 2 cups partially thawed green beans or lima beans for the broccoli. Omit cayenne pepper and add 1 chopped fresh tomato at the last minute. Sprinkle with ¼ cup grated Parmesan cheese in place of seeds.

CHAPTER 10

FRUIT

The freezer is an incredible boon to fruit growers and those who frequent roadside markets and farmers' markets. When fruit is bursting out all over, there's your freezer to accommodate the abundance. There are an enormous number of ways to store fruit in the freezer and use frozen fruit from your freezer: in wet packs and dry packs and juices and jams and purees, in sauces and shakes and nectars, in creams and coolers and crisps, in ice creams and ices. There are instructions and recipes for all these diverse ways of using fruit in this chapter and several others in the book.

Let the recipes spark your imagination. Fruit lends itself to every variety of blending—mixing with cream, with syrup, with ice, with milk, with eggs, with yogurt and buttermilk, with other fruits, with meats and poultry, with lemon and honey and mint. If you have fruit in your freezer, you will inevitably find yourself experimenting with it.

Buy fruit or pick fruit for freezing that is perfectly ripe. Avoid fruit that is cheap because it's slightly over the hill. Fruit that is already soft will not fare well in the freezer, where fruit inevitably softens to some extent anyway. The diminished firmness of overripe fruit will become mush in the freezer.

Don't buy fruit in large quantities unless you have the time to freeze it immediately. Fruit that waits for you on the countertop or in the refrigerator is losing nutrients and quality. Your goal should be to freeze it very quickly while it is very fresh.

First, sort and wash your fruit gently. It's best to use cold—even iced—water, particularly when you wash berries. If you wash berries in warm water, they may soften and even bleed juice. Cold water, on the other hand, firms berries and fruit. Don't let your fruit stand in water after washing; it will leach out vitamins you'd much rather it held onto. Drain the fruit thoroughly.

Peel the fruit, trim it, pit it and slice it according to the directions given in Preparing Fruit for the Freezer, later in this chapter. Cut large fruit into uniform pieces. Use any bruised fruit for crushed or pureed packs. Pack the fruit dry or with honey.

To tray-freeze fruit before freezing it permanently, arrange the fruit in a single layer. Leave space between the pieces so they do not freeze in clumps. Cover the tray. Tray-freezing of berries and other small fruit pieces usually takes between two and four hours. If you transfer the frozen fruit to a freezer bag, be sure to press out all the air with your hands before sealing it.

Remember not to overload the freezer in your zeal to dispatch your bounty. The rule of thumb to use is to freeze only 2 or 3 pounds of food at one time for each cubic foot of freezer capacity.

Fruit packed dry or in honey or syrup will last up to a year in the freezer. Thaw fruit in the refrigerator in its sealed container and eat it while it is still icy or just barely thawed. A 1-pint package of fruit packed dry will be ready to eat after six to eight hours in the refrigerator. Honey- or syrup-packed fruit takes eight to ten hours per pint. When you use your thawed fruit in a recipe you then intend to freeze, you must adjust the time you keep that recipe in the freezer to compensate for the time that the fruit has already been frozen. That is, if you prepare a recipe with fruit that you've already frozen for two months you will then have to subtract two months from the maximum time you can freeze that recipe.

Tray-freezing is a good way to keep fruits and vegetables from sticking together when frozen. Place the pieces on a tray, with a small space between each piece. When firm, transfer the pieces to a freezer container.

APPROXIMATE YIELD
OF FROZEN FRUITS FROM FRESH

FRUIT	FRESH	FROZEN
Apples	1 bushel 1 box 1¼ to 1½ pounds	32 to 40 pints 29 to 35 pints 1 pint
Apricots	1 bushel 1 crate ⅔ to 1 pound	60 to 70 pints 28 to 33 pints 1 pint
Avocados	4 avocados	1 pint
Bananas and Plantains	5 to 6 fruits	1 pint
Berries	1 crate 1⅓ to 1½ pints	32 to 36 pints 1 pint
Cherries	1 bushel 1¼ to 1½ pounds	36 to 44 pints 1 pint
Coconut	1 to 1¼ coconuts	1 pint
Currants	2 quarts ¾ pound	4 pints 1 pint
Figs	¾ to 1½ pounds	1 pint
Grapefruit	2 pounds	1 pint
Grapes	2 pounds	1 pint
Kiwi fruit	6 kiwi fruits	1 pint
Lemons	11 lemons	1 pint
Limes	15 limes	1 pint
Mangoes	2 to 3 medium mangoes	1 pint
Melons	1 to 1½ pounds	1 pint
Nectarines	1 to 1½ pounds	1 pint
Oranges	6 oranges	1 pint
Peaches	1 bushel 1 to 1½ pounds	32 to 48 pints 1 pint
Pears	1 bushel 1 to 1¼ pounds	40 to 50 pints 1 pint

FRUIT	FRESH	FROZEN
Persimmons	1½ to 1¾ pounds	1 pint
Pineapple	5 pounds	4 pints
Plums and Prunes	1 bushel 1 crate 1 to 1½ pounds	38 to 56 pints 13 to 20 pints 1 pint
Quince	1¼ to 1½ pounds	1 pint
Rhubarb	15 pounds 1 pound	15 to 18 pints 1 pint
Strawberries	24 quarts 1 to 2 quarts	24-30 pints 1 pint

Packing Fruit for the Freezer

Fruit oozes juice when you freeze it. Blueberries, raspberries and cranberries are the only exceptions to this rule. Without juice in the cells, the cell walls of the fruit collapse and the texture deteriorates right along with them. As a result, the fruit turns mushy.

Dry Pack

If you don't at all mind if the fruit gets mushy when frozen because you only intend to cook with it, you can freeze it as is. That's called a dry pack. To freeze fruit in a dry pack, all you do is pack it into a container and freeze it. The fruit will freeze together in a solid block. If you'd rather it didn't, tray-freeze the fruit as described on page 247. You do not have to leave headspace when you dry pack fruit for the freezer because the space around the individual pieces will suffice for expansion. Take out the frozen fruit by the handful to use in making muffins or pancakes or to sprinkle on cereal.

Wet Packs with Unsweetened Liquid

The second possibility is to pack the fruit in unsweetened liquid. We recommend using either water or apple juice. This method is useful for people who prefer not to sweeten their food, or whose diets do not allow it. Add the fruit to a container and shake it to pack the fruit down a bit. Cover the fruit with the liquid, pressing the fruit down under it if necessary. Crumple some plastic wrap or wax paper and fit it just under the surface of the liquid to keep buoyant fruit submerged so it will not darken.

Honey Pack

A third option is to drizzle honey over the fruit. Honey improves the flavor, texture, color and keeping qualities of the fruit you freeze. As the fruit freezes, the juices combine with the honey to produce a syrup. The fruit absorbs the syrup and stays firm. Honey packs work best with fruit that is naturally juicy, like peaches. To prepare it, cut the fruit, place it in a bowl and mix it gently with honey until the juices are drawn out of the fruit and the honey is dissolved. Pack the fruit into rigid containers. The syrup should completely cover the fruit; if it doesn't, add a little water. Because it has only a minimal amount of liquid, fruit frozen in a honey syrup is best when you cook with it.

Honey Syrup

The fourth and most important option for freezing fruit is honey syrup. It's the method to use for the fruit you want to serve uncooked for dessert, because it does the best job of keeping the fruit firm. This is the option to pick when the texture of your frozen fruit is really important.

Always use a light, mild-flavored honey like clover, locust or alfalfa, whose flavor will not dominate the flavor of the fruit. A dark honey like buckwheat has too strong a flavor. Most recipes will suggest you use sugar when freezing fruit. But you can substitute honey in your favorite recipes by using half the quantity specified for sugar.

For a thin syrup, dissolve 1 cup of honey in 3 cups of boiling water. Once the honey has dissolved, do not boil any longer. For a medium syrup, dissolve 2 cups of honey in 2 cups of boiling water. Chill the syrup before adding it to fruit. You can of course experiment with these proportions, accommodating the sweetness of the syrup to the sweetness of the fruit. If the fruit is well packed into the container, ½ cup of syrup will be enough for a pint container. One cup of syrup will be enough for a quart container.

When you pack fruit with a honey syrup, first fill the container with the fruit and shake it to settle the contents. Pour the syrup over the fruit, pressing the fruit down. The syrup should completely cover the fruit. Cover the top of the fruit with crumpled up plastic wrap, foil or wax paper. Cover the container.

When Fruit Discolors

You do not have to be a helpless bystander while your peaches, pears, apricots, sweet cherries and figs turn brown with impunity. Just as there are ways to control their tendency to soften in the freezer, there are ways to control the tannins that darken the flesh of fruits when they are exposed to the air. First, freeze

FRUIT VARIETIES TO FREEZE

Here's a sampling of fruit varieties that can be frozen. This is not an exhaustive list, so if you don't find your favorite listed, experiment with a small quantity to test its freezing quality.

Apples
 Baldwin, Golden Delicious,
 Gravenstein, Jonathan, Lodi,
 McIntosh, Northern Spy,
 Winesap, Yellow Newton Pippin
 Not recommended: Red
 Delicious

Apricots
 Chinese Golden, Moorpark,
 Stark Earli-Orange, Wilson
 Delicious

Avocados
 Bacon, Fuerte, Hass

Berries, miscellaneous
 buffalo berry, cranberry,
 elderberry, huckleberry,
 mulberry

Blackberries
 boysenberry, Darrow,
 loganberry, youngberry

Blueberries
 Berkeley, Bluecrop, Brightblue,
 Delite, Herbert, Ivanhoe, Jersey

Cherries, sour or tart
 English Morello, Montmorency,
 North Star

Cherries, sweet
 Bing, Black Tartarian, Emperor
 Francis, Lambert, Napoleon
 (Royal Ann), Schmidt, Stark
 Gold, Van, Windsor

Currants
 Fay, Perfection, Red Cross, Red
 Lake, White Imperial, Wilder

Melons
 cantaloupe (muskmelon),
 Casaba, Crenshaw, honeydew,
 Persian

Nectarines
 Mericrest, Stark Sunburst,
 SunGlo

Peaches
 Belle of Georgia, Cresthaven,
 Early White Giant, Elberta,
 Habrite, Halehaven, J. H. Hale,
 Madison, Red Haven, Reliance,
 Suwannee, Veteran

Persimmons
 Fuyu, Giant Fuyu, Hachiya,
 Japanese, Tamopan, Tanenashi

Plums and Prunes
 Stanley, Valo

Raspberries
 Albritton, Canby Thornless,
 Cumberland, Cyclone, Dixieland,
 Dorman Red, Earlidawn,
 Heritage Everbearing, Sparkle

Rhubarb
 McDonald, New Valentine,
 Victoria

Strawberries
 Cyclone, Darrow, Earliglow, Fort
 Laramie, Honeoye, Hood,
 Midland, Midway, Ozark
 Beauty, Pocahontas, Redchief,
 Red Star, Rockhill, Sparkle,
 Surecrop, Tennessee Beauty,
 Tioga, Totem, Tribute, Tristar,
 Trumpeter

only mature fruit, because immature fruit is higher in tannins. Prepare it quickly and use it as soon as possible after you thaw it in order to minimize its exposure to the air.

For dry packs: Dip the fruit pieces in a gallon of water containing either ¼ cup of lemon juice or 2 tablespoons of rose hips concentrate, or add ½ teaspoon of ascorbic acid powder, crystals or crushed tablets to 1 quart of water. Ascorbic acid, in any of several forms, will help prevent discoloration. Use it sparingly because of its sharp taste.

For fruit that is being packed with liquid or syrup: Add 1 tablespoon of lemon juice or rose hips concentrate to 1 quart of the pure honey or honey syrup before you mix the syrup with the fruit. Check the frozen fruit after a few days to see if you have used enough of the anti-darkening agent to keep the fruit from browning.

Make up fillings for fruit pies in multiple batches. Mold each one into a pie shape by freezing it in a freezer bag inside a foil-lined pie plate. When the fruit is solid, remove the plate. When it's time to make pie, reach into the freezer for the frozen filling, remove it from the bag, pop it in a pie shell and as it thaws in the oven it makes pie at the same time.

Preparing Fruit for the Freezer

Apples

Apples freeze well, but they are much better canned. Freezing tends to soften and darken them. If you want to freeze apples, wash, core and peel them. Cut them in halves, quarters or slices. Use a dry pack, honey pack or honey syrup with added lemon juice.

Freeze the apple peels only if they are unsprayed. Mince the peels in a blender or food processor, then freeze with any liquid that has accumulated. Add the bits of peel to muffins and cakes for added flavor, fiber and nutrients. (See recipe for Applesauce, page 263.)

Apricots

Apricots freeze well, but they are generally better canned or dried. Freeze them whole, halved or in quarters, and with or without the pits. The pits add flavor if you leave them in the fruit or add a few to the container. (Loosen the skin by dipping the fruit in boiling water for 15 to 20 seconds, then plunging the fruit into a pan of ice water. The skin should slide off easily.) Use a honey pack or a thin or medium honey syrup with added lemon juice or rose hips concentrate.

Avocados

Avocados freeze best as a puree. Wash, peel and cut the avocados in half. To remove a stubborn pit, place the avocado half on a cutting board with the pit facing you. Using a heavy knife, carefully whack the pit with the sharp edge. The pit should lift right out. Mash the pulp. Add 1 tablespoon lemon juice per pint. Dry pack.

Bananas and Plantains

Freeze leftover bananas that have turned brown and use them for cakes and breads. Whole bananas will hold their shape, even when sliced, but will darken when frozen. Freeze the fruit in the skin or peel it. If you freeze bananas with the peels on, allow them to thaw for fifteen minutes before peeling. Use a blender, food processor or fork to mash the bananas if you freeze them as a puree. Add 1 tablespoon of lemon juice to each cup of mashed fruit. Dry pack.

For a special treat in summer, cut ripe bananas in thirds, roll them in a mixture of carob powder and orange juice, then in unsweetened coconut or chopped nuts. Insert a Popsicle stick in each banana, tray-freeze, then transfer to a plastic freezer bag. Eat frozen. Frozen bananas also add smooth, thick texture to ice cream made in the blender or food processor.

Berries

This category encompasses blackberries, blueberries, boysenberries, cranberries, dewberries, elderberries, huckleberries, loganberries, mulberries, raspberries and youngberries. They all freeze well, but are best used for jellies, jams, sauces and drinks. Do not wash unsprayed berries before freezing. Sort and wash all other berries in cold water, discarding any that are discolored, wrinkled or damaged. You can freeze them individually on a cookie sheet, then transfer them to a freezer bag or container, or freeze them in a honey pack, in honey syrup or as a puree.

Cherries

Both sweet and sour cherries freeze well. Sort, wash in cold water and pit the cherries. If you leave the pit in the fruit, it will lend it an almondy flavor, but it will also be harder to remove when the fruit is softer after thawing. Tray-freeze the cherries and transfer them to plastic freezer bags, or freeze them in a honey pack or honey syrup pack with added lemon juice or rose hips concentrate.

Freeze the cherries for pie in a dry pack. Freeze the cherries you want to keep whole to serve in fruit salad in a honey pack or a syrup.

Citrus Fruits

Grapefruit, oranges, lemons and limes freeze very well. Peel and remove the membranes and white skin surrounding the sections. Pack dry, placing a double layer of wax paper between layers for easy removal, or freeze in a honey pack or honey syrup.

Coconut

Fresh coconut meat freezes very well. Pierce two or three of the eyes of the coconut with an ice pick. Drain out the liquid, and drink it or use it for cooking. To open the coconut, hold the fruit in the palm of your hand, with the stem facing upward. Using a small hammer, tap all around the middle, turning the coconut after each tap. Keep tapping and turning the coconut until it splits. Or, bake it in a 350°F oven until the shell starts to crack, about twenty minutes. When it cools, break the shell open with a hammer. Cut out the meat and shred it. Pack the meat in its own milk, or toast it in a 350°F oven for 20 to 30 minutes. You can also freeze coconut in large pieces, or freeze it after grating it in the blender or with a meat chopper.

Currants

The large varieties are best for freezing. Remove the stems and wash the currants. Freeze them in a dry pack or a honey pack.

Figs

Figs freeze very well. To prepare them, wash the fruit and remove the stems (be gentle; figs bruise easily). If the skin is thin, leave it on; otherwise, peel the figs. Keep them whole, or cut them in half and scoop out the seeds with a spoon. To freeze figs in even smaller pieces, quarter or slice the halves. Pack in a honey pack or honey syrup. You can also remove the seeds and crush the figs in a food processor or blender and freeze as a puree. Pack the puree in honey syrup.

Grapes

Freezing makes grapes slightly soft and limp, but it does not affect their flavor. To prepare grapes for freezing, wash and sort them and remove the stems. Freeze seedless grapes whole but pit grapes with seeds. Tray-freeze them, then transfer to plastic freezer bags. Or freeze them in a honey syrup. Eat grapes or use them in salads while they are still slightly frozen and can still hold their shape. Frozen grapes peel much more readily than fresh ones. Just dip them in cool water and the skin will slip right off.

Guavas

These tropical fruits freeze adequately, but with considerable change in texture. Wash, peel and halve the fruits and remove the seeds. Puree in a food processor or blender or mash with a fork. Freeze the puree in a dry pack or honey pack or freeze halves in thin honey syrup.

Kiwi Fruit

Kiwi fruit freezes very well. Peel and cut it into slices and freeze it with a honey syrup. Don't dry pack kiwi fruit; when frozen without liquid it becomes astringent.

Mangoes

Mangoes freeze well. Serve them partially frozen, while they still retain their shape. To prepare them for the freezer, wash, peel and cut off the stem end. Trim away any meat near the seed. Cut the fruit into ¼ inch slices. Freeze in a honey pack or honey syrup, with lemon juice added to prevent darkening.

Melons

Melons, including cantaloupe, Casaba, Crenshaw, honeydew and Persian, freeze well. The only kind that doesn't is watermelon, which becomes mushy and should be either eaten while still partially frozen or used in purees or drinks. The other melons hold their shape well. To freeze a melon, peel it and remove the seeds, and cut the fruit into slices, cubes or balls. Freeze in a dry pack, honey pack or honey syrup.

Nectarines

Nectarines freeze fairly well, but are better canned, where they do not require peeling. To freeze them, wash, and peel the skins with a small sharp knife. The peeling, while time-consuming, is necessary because the skins toughen in the freezer. Halve the fruit and remove the pits. Leave it in halves, or slice into quarters or smaller slices. Immediately after cutting, drop the pieces into lemon juice to prevent darkening during freezing. Freeze in a honey pack or honey syrup.

Peaches

Peaches freeze fairly well but are better canned. Wash and peel by dropping the fruit into boiling water for 30 seconds, then draining and plunging them into ice water. The skins should pull off easily. Halve the fruit and remove the pits. Leave it in halves or slice into quarters or smaller slices. Immediately after cutting, drop the pieces into lemon juice to prevent darkening during freezing. Freeze in a honey pack or honey syrup.

Pears

Pears freeze poorly. They are much better canned or dried or used for jellies and jams. If you do freeze them, wash, core and peel them and halve or slice the fruit directly into lemon juice to prevent darkening. Drain them well, then freeze in a honey pack or honey syrup.

Persimmons

In China and Japan persimmons are always dried, but in this country they are most commonly frozen. Wash, peel and slice the fruit, remove the seeds and freeze the fruit in a dry pack or honey pack with a little added lemon juice. Or, puree the slices in a blender or food processor. Add 2 tablespoons of lemon juice per pint, then freeze. You can also freeze persimmons whole and unskinned. Serve them partially thawed over ice cream.

Pineapple

Pineapple freezes very well, but do not freeze overripe fruit. There are several ways to peel and cut a pineapple, depending on how you want to use it. You can

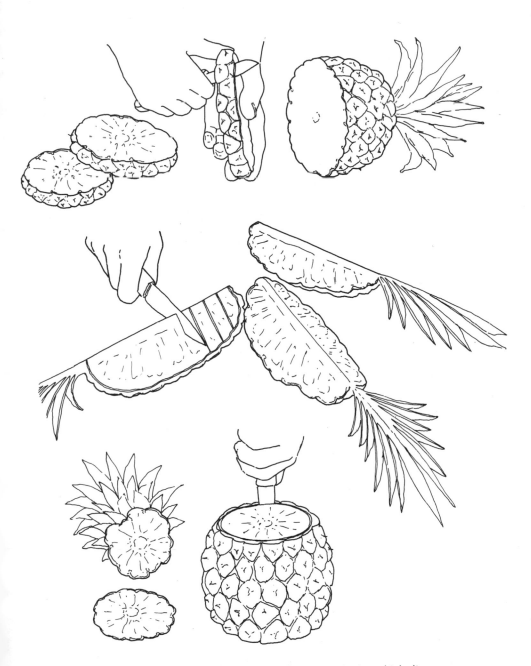

Pineapple can be cut several ways: Top, cut the fruit crosswise into thick slices, then remove the rind; center, cut the pineapple into quarters lengthwise, then free the flesh from the rind and cut into chunks; or bottom, remove the top and bottom, run the knife along the outside edge of the fruit and remove the rind to yield a whole chunk.

use a professional pineapple peeler, or if you want slices, simply cut ½-inch slices across the pineapple, then peel the slices. To make pineapple rings from the slices, remove the fibrous core with a small round cookie cutter, or with a knife. Or you can cut the pineapple in quarters lengthwise, slice along the outside of the fruit to remove the rind, cut away the core and then cut the wedges into chunks. If you want to freeze a whole pineapple, cut off its top and bottom, then run your knife all around the outside edge of the fruit and simply lift out the contents. The best ways to cut pineapple for freezing are in rings or chunks. Pack it in its own juice, in a honey pack or a thin honey syrup.

Plums and Prunes

Plums and prunes do not freeze well. They become very soft when thawed and are better canned. But if you do wish to freeze them, wash and pit the fruits. Leave them whole, or cut them into halves or quarters. Pack dry, in a honey pack or in honey syrup.

To puree plums for freezing, place pitted plums, slightly crushed, in a pot and add ½ cup of water for each quart of fruit. Bring the water to a slow boil and cook for two minutes. Add an additional small amount of water, if necessary. Drain and allow the fruit to cool. Puree in a food mill, sieve or blender. Freeze in a honey pack or honey syrup, with lemon juice to taste.

Pomegranates

Pomegranates freeze well. To prepare one, cut halfway into the fruit, and use the cut to gently break it open. Pop out the seeds, being careful not to bruise them. Freeze the seeds. To use frozen seeds, sprinkle them while still frozen onto fruit salads, applesauce or pear sauce. The seeds thaw very quickly.

Quince

Quince freezes well enough to permit its use in jellies, jams and sauces. To prepare it, peel, quarter and core the fruit, removing the hard pieces around the core. Cut it into chunks. Steam blanch the quince for four minutes or water blanch it for two minutes (see page 198). Drain and immediately plunge the chunks into ice water to stop the cooking process. Drain very thoroughly, tray-freeze, then transfer to freezer bags.

Rhubarb

Early spring rhubarb freezes very well. Prepare only stalks that are red and crisp. Remove the leaves and cut off any woody ends. Wash the stalks thoroughly under cold water and cut into ½-inch to 1-inch chunks. Tray-freeze the chunks, then transfer to plastic freezer bags. Or, blanch the chunks over steam or in water for 1½ to 2 minutes (see page 198). Drain and immediately plunge into ice water to stop the cooking process. Pack in honey syrup.

To make a puree, combine 1 pound of rhubarb, ⅓ cup honey and the grated rind of half a lemon in a pot. Cover and cook over low heat until rhubarb is soft. Stir occasionally to prevent scorching. Puree the mixture in a food mill, blender or food processor. Pack and freeze.

Strawberries

Strawberries freeze very well. Wash and remove the hulls, and let the berries dry. Freeze small berries whole; halve, quarter or slice large berries. Tray-freeze, then transfer to plastic freezer bags, or freeze the berries in a honey pack or honey syrup. Serve strawberries while still a little frozen, while they retain their shape.

To puree your strawberries, put them through a food mill or sieve. Don't use a blender, which will add too much air to the puree. Freeze the puree with honey to taste.

HEADSPACE FOR FRUIT

Unless you are tray-freezing, freeze fruit in rigid plastic containers, freezer bags or freezer-safe glass jars with straight or flaring sides. Don't freeze fruit in glass jars with narrow tops.

	Pint	Quart
Fruit packed in liquid	1 inch	1 inch
Fruit packed dry	1 inch	1½ inches
Fruit juice	1 inch	1½ inches
Jam	1 inch	1½ inches
Puree	1 inch	1½ inches

Freezing Fruit Juice

Homemade fruit juice is in a class by itself. While cooked or canned fruit juice is pleasant, it's pretty unremarkable. Grape juice from a fruit press, on the other hand, is sweet and a little cloudy, thick without being syrupy. It holds the essence of the freshly picked grapes. It's as clear a distinction as that between bottled apple juice and the fresh, unfermented cider that comes straight from a cider press. Juicing part of the harvest is an excellent way to use fruit that's a little too bruised to freeze or can. Juice is juice, after all, even if it's locked inside a less-than-perfect skin. As long as the fruit is really ripe, juice it.

Freezing is the simplest way to preserve juice, and fortunately it is also the best way. Juice that has been frozen tastes uncooked and retains almost all its vitamin content. That is the most important consideration in juice making: holding on to the fresh, fine flavor of the fruit.

Use your fruit juice as more than a breakfast drink. You can blend or process frozen juice until it is as thick as soft sherbet, then add sour cream or whipped cream or some fresh fruit on top. Or mix the fruit juice with an equal quantity of yogurt or buttermilk to thin it a little. It will still be thick and delicious.

Making the Juice

Making juice is easy. Just simmer the fruit in its own juice or a little water to prevent scorching. In the case of cranberries, plums and grapes, use equal quantities of fruit and water, stirring occasionally and mashing the fruit a little to encourage it to open. Add a little honey or lemon juice to taste. Use only

To strain juice from fruit, ladle the crushed fruit into a piece of wet cheesecloth. Gather the ends together and twist until no more juice flows from the fruit.

stainless steel, glass or enameled steel pots, pans and knives. Don't use zinc or aluminum because these metals can leach into the juice when they interact with the acidic fruit.

As soon as the fruit is soft, remove it from the heat. Heat extraction of juice should be as brief as possible, to avoid having the heat adversely affect the flavor of the juice. Once the fruit is soft, the heat will have also stopped the enzymes and eliminated the microorganisms, and it's time to strain the juice. Strain it through a jelly bag, a food mill, a colander or through a couple layers of wet cheesecloth. Save the pulp and freeze it to use in sauces or desserts. If you have cooked your fruit to extract its juice, no further heat treating is necessary. Pour it into freezer jars, leave 1 inch of headspace for pints and 1½ inches for quarts, cool it down quickly, label and freeze for up to twelve months. When you thaw it, drink it within two weeks. It will thaw in the refrigerator in one to two days per quart.

Pasteurizing the Juice

You needn't pasteurize juice that you have made by the heat extraction method described above. Grape juice, however, should always be pasteurized, in order to control its tendency to become wine. You should also pasteurize apple and pear juice that you intend to keep for longer than six months. You should always pasteurize any other juices that are cold extracted, that is, fruits which you have made into juice without heat.

Pasteurizing fruit juice is the equivalent of blanching vegetables. It's best to pasteurize juice in quantities small enough to heat and cool rapidly, so that it retains a fresh-squeezed quality. Stir occasionally to keep the juice from scorching. Using a candy or jelly thermometer, wait for the temperature in the juice to reach 190°F, then remove the juice from the heat and pour into freezer containers. Pour it very slowly so you will incorporate as little air as possible into the juice. The temperature of the juice should not drop below 185°F before it gets in its jar. If the temperature does drop because you've made a large quantity, reheat the cooled part back to 190°F.

Making Frozen Juice Concentrates

You cannot cook down fruit juice to concentrate it in the way you can cook down stock. Fruit juice loses its fresh taste if you cook it long enough to reduce it. It is possible to make frozen fruit juice concentrate at home, but it's a long process. It's probably best to leave concentrates to commercial juice producers. But, if your desire to juice and freeze your harvest is strong, and your freezer space is too limited for any other storage method, here's how to make your own concentrate. First, fill a narrow-necked container three-quarters full of juice. Cap the container and freeze it upside down. When it's frozen, place the container, uncapped, upside down in a glass pitcher. The thawing juice will drip its sweet essence first and its water later. Taste the drips often. When they are no longer sweet, extract the container from the glass pitcher, let the ice thaw out and drink it. Pour the concentrate from the pitcher into the narrow-necked container, freeze it, thaw it in the same way to extract more water. Then repeat the process a third time. At long last, you will have fruit juice concentrate. Pour the concentrate into small freezer containers. To reconstitute it, add three parts water to one part concentrate.

Fruit Purees

Purees are a terrific way to freeze fruit. Pureeing fruit is an easy way to get rid of seeds, pits and slices—you simply strain the fruit, cooked or uncooked, through a sieve. Fruit purees make excellent thickeners. If you puree apples and pears, for instance, you can use them instead of custard to line a tart. The base will be both deliciously fruit-flavored and lower in fat. Fruit purees also make colorful and delectable toppings for a wide range of desserts, and can be delightful fillings for cakes. Of course, they are perfect baby foods, too.

Freezer Jam

You can also freeze in the form of jam some of the juicy fruits and berries you harvest or buy. When you freeze jam you can make it with uncooked fruit and retain the kind of fresh flavor that is exactly what jam is meant to be. Use fruit, agar flakes and a little honey instead of fruit, pectin and a lot of sugar and your jam will have 230 calories a cup, rather than 390 calories a cup.

Pectin, which is the usual vehicle for binding up the fruit juice in jam, requires a lot of sugar in order to work. Agar, which is a seaweed derivative available at natural food stores, does the same job with one-twelfth the amount of sweetening. The Rodale Test Kitchen staff discovered that agar flakes—the most convenient of agar's several forms—mixed with a modest amount of honey made excellent jam.

To make freezer jam (see the recipe on page 287), choose fruits that are very ripe and free of bruises and mold. Berries work best, followed by stone fruits like cherries, peaches, apricots and plums. Try mixing up fruits for interesting flavors. Raspberries and cherries, for instance, make memorable jam together. Hard fruits, like apples and pears, however, do not. They're not juicy enough to set up with agar. Jam can be frozen for nine months, and thaws in the refrigerator at a rate of eight hours per half pint.

APPLESAUCE

You can freeze this applesauce for 9 months.

4 pounds apples, washed, cored and quartered	honey or maple syrup, to taste
1 cup water	

If you are planning to mash the cooked apples by hand, remove the skins before cooking. If you are going to put the cooked apples through a sieve, food mill, blender or food processor, leave the skins on to add fiber and nutrients and to slightly increase the yield.

In a large pot, combine apples and water, bring to a boil and reduce heat. Cover and cook until the apples are tender, about 10 minutes.

Drain the liquid from the cooked apples and reserve. Mash apples with a potato masher, fork or spoon, or use a sieve, food mill, blender or food processor.

Add honey or maple syrup. For a thinner consistency, add some of the reserved liquid.

Freeze in serving-size portions.

YIELD: 4 to 5 cups

NOTE: The following varieties make especially good applesauce because they cook down and soften quickly: Cortland, Duchess, Gravenstein, Greening, Macoun, McIntosh, Northern Spy, Red June, Rome Beauty, Stayman and Yellow Transparent. For a well-balanced flavor, mix tart and sweeter varieties. Avoid using Red Delicious, which cooks down to sauce that is both poorly colored and poorly flavored.

VARIATIONS: Add cinnamon, allspice or nutmeg. Add raisins. Add lemon juice or lemon peel.

Mix the applesauce with an equal quantity of another fruit puree or fruit sauce. Try an apple-peach combination, or apple-apricot, or apple-plum, all of which have an attractive color. Or try an apple-pear combination.

CINNAMON-APPLE SAUTÉ

This is an unusual and wonderful side dish to serve with poultry or pork.

1 tablespoon corn or safflower oil	½ teaspoon ground cinnamon
3 tablespoons butter	½ cup broken pecan pieces
4 cups frozen apple rings or slices (baking variety), thawed enough to separate pieces	

In a large skillet, heat oil and butter until foamy.

Add apple rings and sauté briskly over moderately high heat. When apples are hot, after about 8 minutes, sprinkle cinnamon on apples and toss in pecans.

Toss and cook 3 to 4 minutes longer. Serve warm.

YIELD: 4 to 5 servings

VARIATIONS: Use peaches or pecans instead of apples.

APPLE-PECAN PIE FILLING

This rich pie filling is great baked in a double pie crust. Serve warm with a scoop of vanilla ice cream for an excellent autumn dessert. The filling can be made ahead of time, frozen, thawed and popped into a shell and baked for a treat any time of the year. It freezes for 6 months.

4½ pounds apples, about 14 medium-size apples	½ teaspoon ground cardamom
2 cups chopped pecans	2 teaspoons grated orange peel
1 cup raisins	
2 teaspoons ground cinnamon	⅓ cup whole wheat pastry flour
¼–½ cup honey, depending on tartness of the apples	2 tablespoons lemon juice

Peel, core and slice apples. Place in large bowl. Add pecans, raisins, cinnamon, honey, cardamom, orange peel and flour. Sprinkle with lemon juice and mix well.

At this point, you may freeze in plastic bags, preferably doubled. Or, place filling in prepared whole wheat pie shell (page 329) and bake in a 375°F oven for 40 to 45 minutes.

Defrost frozen filling before using it.

YIELD: enough filling for two 9-inch pies

VARIATION: Substitute 1 cup chopped dates for the raisins, and 1 teaspoon mace for the cardamom; eliminate the orange peel and prepare as directed above.

∗GUACAMOLE∗

This creamy dip holds its nice green color when frozen, even though it contains no lemon juice! You can freeze it for 6 months.

2 ripe avocados
2 tablespoons finely chopped onions
1 small tomato, finely chopped
1 tablespoon finely chopped jalapeño pepper
1 tablespoon chopped fresh coriander, optional

1 small clove garlic, minced
freshly ground black pepper, to taste
2 tablespoons sour cream, optional

Peel, pit and mash avocados. Mix in onions, tomatoes, peppers, coriander and garlic. Add ground pepper.

Serve immediately or freeze in a covered container. Thaw dip before serving.

Mix in 2 tablespoons sour cream just before serving, if desired.

YIELD: 2 cups

BANANA FRUIT SHAKE

1 small ripe banana,
     peeled and frozen
¾ cup milk
1 tablespoon maple syrup
     or honey
½ cup frozen strawberries,
     raspberries or sweet
     cherries
	¼ teaspoon vanilla extract
	     (use almond extract
	     with cherries)
	     freshly grated nutmeg

Place frozen banana, milk, maple syrup, strawberries and vanilla in a blender. Process on low to medium speed until smooth.

Place blended drink in two chilled glasses. Sprinkle with nutmeg. Serve immediately.

YIELD: 2 servings

VARIATION: Use yogurt instead of milk and increase maple syrup to 2 tablespoons.

BAKED CARAMEL FROZEN BANANAS

A yummy and nutritious breakfast or dessert.

⅓ cup maple syrup
1 tablespoon honey
3 tablespoons butter
½ teaspoon ground
     cardamom
4 frozen bananas, thawed
     at room temperature
     for 15 minutes
	¼ cup toasted wheat germ
	¼ cup slivered almonds
	1 cup yogurt

To make the caramel, combine maple syrup, honey, butter and cardamom in a medium-size saucepan. Boil gently over medium heat, stirring constantly, for about 10 minutes or until the color darkens slightly. To test for doneness, dip a cold metal spoon into the caramel and then set spoon aside to cool. When cool, the caramel on the spoon should be hard.

Slice frozen bananas lengthwise into an 8 × 8-inch baking dish and pour the hot caramel over the bananas. The caramel will harden when it touches the bananas.

Sprinkle immediately with the wheat germ and almonds.

Bake at 350°F until hot and bubbly, about 15 minutes. Stir to cover bananas with syrup. Serve warm with yogurt for a smooth and creamy contrast to the cardamom.

YIELD: 4 servings

VARIATION: Fresh bananas can be covered with the same syrup and frozen to be baked later. They are best frozen flat in pie pans so that they can be baked without thawing. Add about 15 minutes to the baking time. Can be frozen for 6 months.

BERRY CREAM

This is a lovely, thick drink with the flavor of cheesecake. It's ideal for dessert or as a midafternoon cooler.

2 cups frozen strawberries, blackberries or blueberries
¼ cup yogurt
1½ cups milk
2 teaspoons vanilla extract
⅛–¼ teaspoon ground allspice, optional

2–4 tablespoons honey, depending on sweetness of berries
4 ounces cream cheese, at room temperature

Place berries, yogurt, milk, vanilla, allspice and honey in the blender and puree for 1 minute.

Add cream cheese 1 tablespoon at a time, processing 15 seconds after each addition.

Serve immediately in chilled glasses.

YIELD: 2 to 4 servings

CINNAMON-BERRY NECTAR

Drink this nectar for breakfast, or as a summer cooler mixed with sparkling water. Warmed up, it becomes a sauce to serve with pancakes, waffles, ice cream or yogurt. You can make it with either fresh or frozen berries and freeze it (or refreeze it) for 9 months.

8 cups blackberries or raspberries	¼ cup orange juice
1 cup water	2 teaspoons vanilla extract
½–¾ cup honey	1–1½ teaspoons ground cinnamon

In an 8-quart saucepan, combine berries and water. Bring to a boil over medium heat and cook until the berries release their juices and are soft, 5 to 10 minutes for frozen, 6 to 8 minutes for fresh.

Lower heat and add honey, orange juice, vanilla and cinnamon. Cook 1 minute more.

Remove from heat and puree in a food processor or blender. Cool. If there are too many seeds for your taste, strain through a fine mesh sieve.

YIELD: 4⅔ cups

VARIATIONS: To make a cool summer party drink, place the nectar in a glass and add 4 ounces of mineral water or sparkling water, or use 6 tablespoons of nectar per one 23-ounce bottle of sparkling or mineral water and serve in wine glasses.

MAINE BLUEBERRY PIES

Individual deep-dish pies made with frozen berries. There is no pastry under the pies . . . only on top. Can be frozen for 4 months.

Pastry

5 tablespoons butter	2 tablespoons ice water
1 cup whole wheat pastry flour	2 teaspoons maple syrup

Filling

3 cups frozen dry pack blueberries	2 teaspoons lemon juice
2 tablespoons cornstarch	dash of ground cinnamon
¼ cup honey	2 teaspoons butter
milk to brush crust	

To prepare pastry, cut butter into flour with a pastry blender or two knives, until butter pieces are the size of small peas. Stir in ice water and maple syrup with a fork. Dough will form a soft ball.

Place in plastic wrap or wax paper and refrigerate 30 minutes.

To prepare filling, place blueberries in a medium-size bowl and toss with cornstarch. Divide the berries among four ovenproof bowls or custard cups. Ideally, the bowls should have lips.

In a small bowl, mix together honey, lemon juice and cinnamon. Drizzle mixture over blueberries. Dot with butter. Remove dough from refrigerator. Flour dough lightly and roll out between sheets of wax paper until the dough is 11 inches in diameter. Cut four circles, each 1 inch larger than the diameter of the bowls the pies will be baked in.

Brush the edge of each bowl with water. Place one circle of dough on top of each bowl and crimp the crust against the edge of the bowl. Pierce the top with a fork.

Lightly brush crusts with milk and freeze now in freezer bags or bake in 375°F oven for 30 minutes. Cover with foil and bake 15 minutes more, or until fruit bubbles.

Remove from oven. Serve warm with ice cream, whipped cream or light cream.

To bake a frozen pie, remove from bag and bake at 375°F for 30 minutes. Cover with foil and bake 30 minutes more, or until fruit bubbles. Serve as above.

YIELD: 4 individual pies

BLUEBERRY-ORANGE SAUCE

This sauce is delicious served over ice cream or pancakes, or mixed with hot or cold yogurt. Use fresh or frozen berries and freeze the sauce for up to 9 months for later use.

    8 cups blueberries
    ¾ cup maple syrup
    ½ cup honey

    2 teaspoons grated orange
        peel

Place blueberries in a food processor, 2 cups at a time. If using a blender, it will be easier to do 1 cup at a time. Chop until chunky. (You may have to stop and scrape the sides of the bowl to evenly chop all of the berries.) Repeat to process all the berries and place in a 4- or 5-quart saucepan.

Cook the chopped berries gently over medium heat until they release their juices, between 5 and 8 minutes.

Increase the heat and bring the berries to a boil. Boil for 3 minutes, stirring continuously to prevent scorching.

Stir in the maple syrup, honey and orange peel. Reduce heat. Continue to stir and simmer until the mixture thickens, about 5 minutes.

YIELD: 4 cups

VARIATIONS: Frozen raspberries, strawberries or sweet cherries can be substituted for the blueberries.

BERRY ICE

Can be frozen for 1 day.

    3 cups frozen raspberries
        or strawberries, or a
        combination of the two

    ¼ cup honey

Puree raspberries and honey in a food processor. Strain and freeze. For smoother texture, allow to thaw slightly, then whip in processor again and refreeze or serve immediately.

YIELD: 3 cups

VARIATION: Add frozen juice concentrates to taste and decrease honey.

BLUEBERRY-MELON ICE

French *sorbets* and Italian *granitas* and ices are all pretty much the same thing. They are simple to make and so refreshing to the palate that they are often served between courses at a lavish dinner, enabling people to get their second wind, gastronomically speaking. Make this ice from frozen fruit, but don't refreeze it.

3 cups frozen honeydew   1 cup frozen blueberries
    chunks or balls

Combine honeydew and blueberries in blender or food processor. Process until smooth.
Serve immediately.

YIELD: 2½ cups

VARIATION: Use raspberries in place of the blueberries.

LEMON-BLACK CHERRY ICE

Can be frozen for 1 day.

3 cups black cherries,   ⅓-½ cup honey
    pitted   1 tablespoon lemon juice
1 cup apple juice

Simmer cherries, apple juice and honey together until cherries are tender, 12 to 15 minutes. Pour through a wire strainer, pressing down on the pulp until it is almost dry. Discard pulp. Stir lemon juice into the berry juice.
Cool and freeze juice in a loaf pan. When almost solid, puree quickly in a food processor or blender, then refreeze. The color improves when reblended.

YIELD: 2 to 2½ cups

VARIATIONS: Substitute sliced plums or blackberries for the black cherries.

CRANBERRY-ALMOND CHUTNEY

This spicy, slightly hot condiment is a fascinating blend of western and eastern ingredients. Serve warm or cool with rice, roast turkey or chicken. Make it with fresh or frozen cranberries, and freeze some to eat later. You can freeze it for 9 months.

2 oranges	1-2 teaspoons red pepper
2 lemons	flakes
2 cloves garlic	3 tablespoons vegetable oil
2 tablespoons coarsely	7 cups cranberries
chopped fresh	1 cup honey
gingerroot	1 cup raisins
1 teaspoon ground	1½ cups coarsely chopped
cinnamon	toasted almonds
2 teaspoons fennel seeds	(page 54)

Using a vegetable peeler, remove the thin outer peel from oranges and lemons. Chop peel into tiny squares. Juice oranges and lemons and reserve the juice.

Mound garlic, gingerroot, cinnamon, fennel seeds and pepper flakes on a cutting board. Chop together until finely minced.

In a covered, 2-quart saucepan, cook the spice mixture in the oil over low heat for 2 minutes.

Add cranberries, orange and lemon peel and juice. Stir well to coat with oil, raise heat to medium, cover and cook, stirring occasionally, until cranberries begin to burst, about 8 minutes for fresh, about 15 minutes for frozen. Uncover and stir in the honey and raisins. Cook just until the cranberries are tender, about 5 more minutes. Stir in the almonds and set aside to cool.

Serve immediately or freeze by ladling chutney into freezer containers.

To use chutney that has been frozen, thaw at room temperature for 6 hours or in the refrigerator for 36 hours before serving.

YIELD: 3 pints

SOUR CHERRY PIE

5 cups dry pack frozen,
  pitted sour cherries,
  thawed and undrained
½ cup honey
3 tablespoons quick-
  cooking tapioca

1 teaspoon grated lemon
  · peel
  whole wheat pastry for
  two-crust 9-inch pie
  (page 329)
1 tablespoon butter

In a large bowl, combine cherries, honey, tapioca and lemon peel. Let stand in a warm place 20 minutes, stirring occasionally.

Line a 9-inch pie plate with half of the dough. Pour cherry mixture into pie pan and dot with butter. Cover with remaining pastry, seal the edges, and prick all over with a fork. Cover pie with foil and bake at 375°F for 30 minutes. Remove foil and bake until crust browns lightly, about 25 minutes.

YIELD: one 9-inch pie

PINK GRAPEFRUIT REFRESHER

This syrup can be frozen for 9 months.

4½ cups pink grapefruit
  juice, about 6 grapefruit
¼ cup chopped grapefruit
  peel

1½ cups honey
3 tablespoons black cherry
  concentrate
6 cups boiling water

In a large bowl, combine grapefruit juice, peel, honey and cherry concentrate. Stir in the boiling water. Cool thoroughly, strain, pour into freezer containers or ice cube trays and freeze.

To serve, thaw completely and combine with an equal amount of cold water or sparkling water. Serve over ice.

YIELD: 2½ quarts syrup, enough for 5 quarts of beverage

CITRUS REFRESHERS

Lemonade and limeade drinks can be prepared while the fresh fruit is in season and frozen for year-round use. This technique pasteurizes the juice for long-term storage while retaining the vitamin C. You can freeze the syrup for these drinks for 9 months.

2 cups fresh lemon juice, about 12 lemons

1 tablespoon grated lemon peel

1 cup plus 2 tablespoons honey

2 quarts boiling water

In a large bowl, combine the lemon juice, peel and honey. Stir in the boiling water. Cool thoroughly before pouring into freezer containers or ice cube trays and freezing.

To use, thaw completely and combine with an equal amount of cold water or sparkling water. Serve over ice.

YIELD: 2½ quarts syrup, enough for 5 quarts of beverage

NOTE: You will get more juice by first letting the fruit come to room temperature, then pressing and rolling the uncut fruit on a counter before cutting and squeezing it.

VARIATIONS: To make ginger lemonade, add 6 tablespoons peeled and minced fresh gingerroot to the lemon juice. Continue as directed above, except strain mixture before freezing.

To make limeade, substitute limes in places of lemons, and proceed as directed.

FIGS FROZEN IN RASPBERRY SAUCE

Figs can be frozen up to 6 months in this elegant sauce.

2 cups fresh or frozen red raspberries

⅓ cup honey

½ cup apple juice

16 fresh figs

1 cup crème fraîche, optional

½ cup fresh or frozen raspberries, optional

Simmer raspberries, honey and apple juice together until raspberries are tender, about 10 minutes. Cool slightly, puree and strain to remove seeds.

Meanwhile, cut stems off figs and peel, if desired. Cut figs into quarters from top to bottom, leaving the bottom intact to form a flower shape. Place figs in container, cover with cooled syrup and freeze.

To serve, simply thaw and serve in small bowls or on top of custard or ice cream. For a fancier and richer dessert, drain sauce from thawed figs and mix crème fraîche into sauce. Pour back over figs and add a few whole raspberries to each portion.

YIELD: about 4 cups

VARIATIONS: Substitute poached pears or fresh peaches for the figs.

SAUCE VERONIQUE
(WHITE GRAPE SAUCE)

This subtly dramatic sauce is simpler to make with grapes that have been frozen, because they are easy to peel.

2 cups Chicken Stock (page 73), thawed if frozen ground nutmeg, to taste ground black pepper, to taste	½ cup heavy cream 1½ cups frozen white seedless grapes 2 tablespoons chopped fresh parsley

Boil stock until reduced to 1 cup. Reduce heat to a simmer and season with nutmeg and pepper. Add cream and simmer until slightly reduced, about 15 minutes.

Meanwhile, run lukewarm water over grapes until the skins split. Slip skins off and discard. Add peeled grapes to stock-cream mixture and cook until heated through. Add parsley and serve immediately while hot, over baked or poached chicken breasts or fish.

YIELD: about 2 cups

VARIATIONS: You can reduce the parsley and add fresh dillweed, chives or chervil.

KIWI SORBET

This unusual sorbet can be frozen for 1 month.

2 cups kiwi fruit puree
  (9 to 12 kiwi fruits),
  thawed if frozen

¾ cup water
⅓ cup very light honey
juice of ½ lemon

Blend all ingredients in a large bowl. Pour into a thin layer in a metal pan (such as a 9 × 13-inch cake pan) or in several ice cube trays, and place in freezer. When firm to the touch, place in a food processor and process until well blended or blend in batches in a blender. Pour back into pan or trays and freeze. Repeat the processing once more just before serving. Serve as a light dessert or as a refresher between courses.

YIELD: 3 cups

NOTE: If you use a dark honey, the sorbet will have a heavy and not-unpleasant honey flavor, but it will overpower the kiwi fruits.

MANGO CHUTNEY

This is a good way to keep mangoes in the freezer or to use frozen mangoes. The chutney can be frozen for 9 months.

3 mangoes, near-ripe but
  still a bit firm
1 teaspoon vegetable oil
¼ cup cider vinegar
¼ cup water
⅓ cup raisins
3 tablespoons light
  molasses
⅓ cup orange juice
4 teaspoons minced fresh
  gingerroot

½ teaspoon minced garlic
1 4-inch cinnamon stick
2 small dried red chili
  peppers
½ teaspoon ground
  coriander
¼ teaspoon ground
  cardamom
½ teaspoon turmeric
2 cups chopped apples, in
  ½-inch cubes

Peel mangoes. Catch the juice while preparing the fruit. Slice the flesh away from the mango in bite-size pieces. Combine the mango, oil, vinegar, water, raisins, molasses, orange juice, gingerroot, garlic, cinnamon, chili peppers, coriander, cardamom and turmeric in a saucepan.

Cook over medium heat, stirring frequently, until the mango is very soft and the mixture somewhat thickened, about 7 minutes.

Stir in the apples and cook until just tender, about 2 minutes more.

Remove cinnamon stick and chilis. Pack the chutney in freezer containers in serving-size portions, usually 1 cup.

YIELD: 3 cups

SPARKLING HONEYDEW COOLER

Serve in tall, chilled glasses garnished with mint, as an after-dinner cooler or light dessert.

2½ cups frozen honeydew melon chunks	1 tablespoon frozen orange juice concentrate
2 egg yolks	1 tablespoon honey
2 tablespoons heavy cream	½–1 teaspoon ground nutmeg, to taste
2 cups sparkling water, approximately	

Place melon, egg yolks, cream, ½ cup of sparkling water, orange juice concentrate, honey and nutmeg in a blender or food processor. Process until smooth, stopping to scrape the sides of the container as necessary. Add remaining sparkling water to taste.

YIELD: 4 servings

VARIATIONS: Substitute 2½ cups frozen kiwi fruit slices or cantaloupe chunks for the honeydew.

FENNEL HONEYDEW ICE

This is the ultimate palate cleanser. It can be frozen for 1 day.

3 cups honeydew balls or
   chunks

½ teaspoon fennel seeds
½ teaspoon lime juice

Process honeydew, fennel seeds and lime juice in a blender until melon is pureed and fennel seeds are crushed, about 1½ minutes.

Freeze for 2 hours, stirring every 15 to 20 minutes to break up ice crystals.

YIELD: 2 cups

MELON SORBET

Some fruit, like cantaloupe and pineapple, is so naturally sweet that additional sweetening is seldom necessary. If you begin with frozen melon no additional freezing is necessary. The sorbet can be frozen for 1 month.

4 cups cantaloupe, honeydew, pineapple or
   watermelon chunks or
   balls, thawed slightly
   if frozen

1 teaspoon lemon juice

Place cantaloupe in a food processor with lemon juice. Process until pureed, and freeze. Remove from freezer, thaw slightly, process again if smoother texture is desired, then refreeze.

YIELD: about 3 cups

VARIATION: For a creamier texture, add 1 sliced frozen banana along with the melon when processing.

PEACH-BERRY NECTAR

Can be frozen for 9 months.

2 cups peaches   1 cup strawberries or
2¾ cups apple juice   raspberries

Puree peaches and ½ cup apple juice in a food processor or blender.
Pour into large bowl.

Puree strawberries and ¾ cup apple juice in a food processor or
blender. Strain mixture to remove seeds, and add to the peach puree.
Stir remaining apple juice into fruit puree.

Serve or freeze. The nectar is delicious while still partially frozen.

YIELD: 4 to 6 servings, 1 quart

VARIATIONS: Substitute apricots for the peaches and add another ½
cup of apple juice.

Red- or golden-fleshed plums can also be used in place of the peaches
(other varieties will be unattractive).

BAKED PEACHES
WITH PECAN STREUSEL TOPPING

4 cups frozen peach slices,   4 teaspoons butter,
   thawed just enough   at room temperature
   to separate slices   4 teaspoons whole wheat
1 cup chopped pecans   flour
¼ cup molasses

Arrange peaches in a single layer in a shallow 9 × 13-inch ovenproof dish.

Work pecans, molasses, butter and flour together in a small bowl
until well mixed. Spoon pecan mixture evenly over peaches.

Bake at 375°F for 20 minutes, or until peaches are hot and bubbly.
Serve warm or cold, plain or a la mode.

YIELD: 4 to 6 servings

VARIATION: Use apple slices instead of peaches, and add ½ teaspoon
ground cinnamon to the pecan mixture.

PEAR SAUCE

You can use this pear sauce like applesauce. It's more elegant than apple-sauce, however, and lends itself to dessert toppings and crepe fillings. It can be frozen for 9 months.

3 pounds pears, washed, cored but not skinned, and cut into 1-inch slices

1 cup water

In a medium-size saucepan, bring pears and water to a boil. Lower the heat and simmer pears until tender, about 5 minutes. Drain and freeze cooking liquid for future use (syrups, nectars, etc.).

Puree the fruit in a food processor or blender until smooth.

Freeze the sauce in pint containers.

YIELD: 3½ to 4 cups

NOTE: The best pears to use for this sauce are those that do not store well when whole, such as Anjou, Bartlett or Bosc.

FRESH PINEAPPLE SAUCE

Sauce can be frozen for 9 months if made with fresh pineapple. If you use frozen pineapple, for every month the pineapple was in the freezer, subtract 1 month from the storage life of the sauce.

2 cups pineapple chunks, thawed if frozen
2 tablespoons maple syrup
¼ teaspoon ground coriander

freshly ground allspice, to taste
freshly grated nutmeg, to taste

Chop pineapple in a blender or food processor until crushed. Place fruit in small bowl. Stir in maple syrup, coriander, allspice and nutmeg.

Serve over hot custard, crepes, French toast or waffles.

YIELD: 1⅓ cups

RHUBARB CHUTNEY

Serve this spicy-hot condiment with cooked brown rice or roasted meats. It is also good spread with cream cheese on toast.

Rhubarb chutney can be prepared with fresh or frozen rhubarb and frozen to use later. Freeze it for up to 9 months.

6 cloves garlic, peeled	1 cup honey
1 1-inch chunk fresh gingerroot, peeled	½–1 teaspoon cayenne pepper
½ teaspoon cumin seeds	2 teaspoons soy sauce
2 oranges	¼ cup raisins
2 lemons	thin slivers of lemon peel or chopped nuts
¼ cup vegetable oil	
6 cups ½-inch-thick slices rhubarb	

Mince garlic, gingerroot and cumin seeds together, keeping them in a round mound, until moist and fine.

Using a vegetable peeler, remove the thin outer peel from the oranges and one of the lemons. Slice this peel into ¼-inch squares. Juice oranges and lemons.

Heat oil over low heat in a large stainless steel or enamel saucepan. Add garlic mixture and simmer gently for 2 minutes. Don't let the garlic brown.

Add rhubarb and fruit peels. Stir well to coat with oil. Cover pan and cook for 4 minutes more if using fresh rhubarb, 10 minutes for frozen.

Stir in the fruit juices and simmer, covered, for 20 to 25 minutes or until the rhubarb is tender.

Add honey, cayenne, soy sauce and raisins. Stir well, and cook, uncovered, until the chutney is as thick as jam, 5 to 15 minutes.

To freeze, cool and pack in freezer boxes. One-cup portions may be the most convenient.

To serve, thaw at room temperature for 6 hours or in the refrigerator for 36 hours. Serve warm or cool, garnished with lemon peel or nuts.

YIELD: 4 to 5 cups

NOTE: To make a Raita (an Indian version of a cold soup), combine ¼ cup chutney with ¾ cup yogurt. Serve as an introduction or accompainment to a meal, or enjoy on its own as a snack.

STRAWBERRY-RHUBARB PIE

Do not refreeze this pie.

3½ cups frozen strawberries,
   partially thawed
3 cups frozen rhubarb
   pieces, partially
   thawed
¾ cup honey

1½ teaspoons lemon juice
¼ cup arrowroot
   whole wheat pastry for
   two-crust 9-inch pie
   (page 329)

In a medium-size bowl, gently stir together the strawberries, rhubarb, honey, lemon juice and arrowroot. Cook over medium heat, stirring until heated through and thickened. Set aside.

Line a 9-inch pie pan with half of the pastry. Spoon filling into pie shell. Carefully top with remaining pastry and crimp or flute edge. Prick top crust with a fork.

Bake at 375°F for 60 to 65 minutes until filling bubbles. If top crust browns too quickly, cover with foil.

YIELD: one 9-inch pie

STRAWBERRY BAVARIAN

This light dessert is based on a molded English custard known as Bavarian Cream. Although the recipe calls for two tablespoons of gelatin, the end result is surprisingly creamy.

2 tablespoons lemon juice
2 cups pureed frozen
   strawberries, thawed
2 tablespoons gelatin
¼ cup water
1 cup milk

2 eggs, separated
¼ cup honey
1½ teaspoons vanilla extract
1 cup heavy cream
¼ cup chopped almonds,
   lightly toasted (page 54)

Stir lemon juice into strawberries.

Sprinkle gelatin into water and stir briefly.

Heat milk in a heavy-bottom pan until hot but not boiling.

Beat egg yolks slightly. Stir ¼ cup hot milk into egg yolks. Return this mixture to remaining milk. Add gelatin and honey to milk.

Stir constantly over medium heat until slightly thickened, 15 to 20 minutes. Do not boil.

Remove from heat and cool to room temperature. When stirred with a wooden spoon, the spoon should leave a path. If it does not, return to heat and continue cooking to thicken mixture further. When thickened, remove from heat, cool to room temperature and fold in strawberries.

Beat egg whites until stiff and fold into strawberry mixture.

Add vanilla to heavy cream and beat until soft peaks form. Fold into strawberry mixture.

Oil a 1½-quart mold. Sprinkle almonds into mold and spoon strawberry cream into mold.

Cover and chill until firm.

To unmold, dip mold briefly into hot water. Unmold onto plate.

Serve topped with additional whipped cream or a custard sauce.

YIELD: 8 servings

NOTE: If you are using strawberries that are frozen whole, partially thaw the berries, then puree them in a blender or food processor.

VARIATION: Substitute raspberries for strawberries.

LUNCH-BOX FRUIT SALAD

Use fresh or frozen fruit in this salad. It can be frozen for 9 months.

1 cup grapes	¼ cup frozen apple juice
1 cup blueberries	concentrate
1 cup sliced peaches	¼ cup water
1 cup strawberries	

Mix fruit thoroughly. Mix apple juice concentrate and water and pour evenly over fruit mixture. Divide evenly into six small freezer containers. Close containers and freeze. A container will thaw in a lunch box in about 3 hours.

YIELD: 6 servings, 4 cups

WINTER-SUMMER FRUIT SALAD

When mixing the fruits, be aware that the juice of dark frozen fruit, such as blackberries, blueberries or raspberries, will stain the white-fleshed fruit like apples.

1 cup thinly sliced fresh apples	1 cup frozen peaches
1 teaspoon lemon juice	1 cup frozen strawberries
1 cup thinly sliced fresh pears	1 cup frozen grapes

Mix apples, lemon juice, pears, peaches, strawberries and grapes. Fruit will be ready to serve when the frozen fruit has begun to defrost, but still holds the shape, about 1¼ to 1½ hours at room temperature. Serve immediately.

YIELD: 4 to 6 servings

VARIATION: Use fresh orange, tangerine, grapefruit with frozen blueberries, blackberries, raspberries.

FRUIT COBBLER

Use frozen sliced peaches, blueberries, blackberries or equal amounts of rhubarb pieces and strawberries to make this treat.

Topping

1 cup whole wheat pastry flour	1 tablespoon honey
½ teaspoon baking soda	3 tablespoons vegetable oil
	½ cup yogurt

Filling

4 cups frozen fruit	2 tablespoons butter
⅓ cup honey	2 tablespoons cornstarch
1 tablespoon lemon juice	¼ cup apple juice

To make the topping, mix together flour and baking soda in a small bowl. With a fork, stir in honey, oil and yogurt just until mixed. Set aside.

To make the filling, in a medium saucepan, heat fruit, honey and lemon juice until hot. Stir in butter and continue heating until butter is melted.

Dissolve cornstarch in apple juice, stir into fruit mixture and cook until thickened.

Place hot fruit filling in a buttered 1½-quart ovenproof casserole. Top fruit with spoonfuls of topping.

Bake at 400°F until light brown and bubbly, 20 to 25 minutes. Serve warm with ice cream.

YIELD: 6 to 8 servings

MIXED FRUIT CRISP

This crunchy crisp is a delicious alternative to plain fruit and cereal.

3 tablespoons melted butter	1 cup granola
2 tablespoons maple syrup	1 cup chopped frozen cranberries, partially thawed
3 cups mixed unsweetened breakfast cereals (corn-flakes, bite-size shredded wheat, puffed corn, rice or millet)	2½ cups chopped frozen apples, partially thawed
	1 cup apple juice

Drizzle melted butter and maple syrup over cereal, add granola and toss until coated.

Gently toss cranberries and apples together and place in a layer on the bottom of an 8 × 8-inch baking dish. Cover the fruit with the cereal mixture. Pour the apple juice over the cereal.

Bake at 350°F until lightly browned, 40 minutes for fresh fruit, 60 minutes for frozen. This casserole can also be prepared ahead of time, refrigerated, and reheated in a 350°F oven for 15 minutes.

YIELD: 8 to 10 servings

VARIATIONS: Substitute a combination of strawberries and rhubarb or peaches and blueberries for the apples and cranberries.

FRUIT SPRITZERS

A variety of fruit juices and berries make this a healthful way to quench your thirst.

1 cup frozen strawberries, raspberries or sliced peaches, partially thawed	½ cup apple juice
	2 ice cubes, cracked
	½–1½ cups cold sparkling water
½ cup orange or pineapple juice	thin lemon or orange slices

Place strawberries, orange juice, apple juice and ice in a blender. Process on medium speed until smooth. Pour into a serving pitcher and stir in sparkling water, using as much as you wish to thin the drink.

Serve in chilled glasses. Cut a slit halfway through each slice of lemon or orange and slip it over the rim of the glass to garnish.

YIELD: 4 servings

FRUIT PUNCH

8 cups frozen strawberries, partially thawed	½ cup lemon juice
	1 cup diced pineapple, thawed if frozen
4 cups frozen Peach-Berry Nectar (page 279) thawed	1⅓ cups water
	⅔ cup honey
1¼ cups orange juice	2 quarts sparkling water

In a blender, puree 6 cups of the strawberries until smooth.

In a large bowl, stir together the strawberry puree, nectar, remaining strawberries, orange juice, lemon juice and pineapple.

In a 1-quart saucepan stir together the water and honey and bring to a boil. Pour the honey-water over the fruit mixture and stir.

Just before serving, gently stir in the sparkling water.

YIELD: 24 servings, 6 quarts

FREEZER JAM

This jam can be frozen for 9 months.

1 tablespoon lemon juice	½ cup cold water
3 cups prepared mashed	3 tablespoons plus 1½
fruit, at room	teaspoons agar flakes
temperature	½ cup mild honey

Stir lemon juice into fruit and set aside. Place water in a small saucepan and stir in agar flakes. Do not stir any further. Wait 1 minute, then bring agar to a simmer over medium-low heat. Once the mixture is simmering, stir for at least 2 minutes or until the agar is completely dissolved. Then stir in the honey. Scrape the sides and bottom of the pot with a rubber spatula.

Pouring with one hand and stirring with the other, add the agar mixture to the fruit. (Do not add the fruit to the agar.) Continue stirring until all ingredients are completely mixed. Taste, and add more honey if desired.

Pour into jars. If you will use the jam within 3 weeks, refrigerate it. Otherwise, freeze it.

YIELD: 4 to 4½ pints

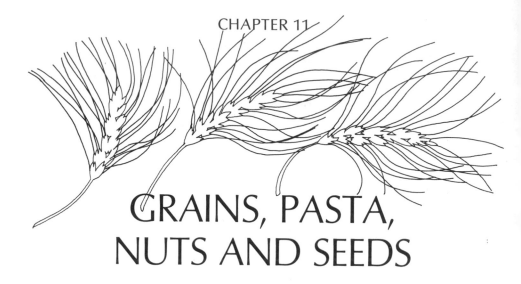

CHAPTER 11

GRAINS, PASTA, NUTS AND SEEDS

Grains are the bedrock of a healthy family's diet. Whether they are cooked or uncooked, the freezer is an excellent repository for them. Not only do whole grains and flours freeze with no loss of quality, they actually benefit from freezing. Freezer temperatures, for one thing, wipe out any trace of the insects or their larvae that may be present in the grains you store. Even if you have no room for long-term grain storage in your freezer, if you freeze your newly acquired grains for as short a period as three or four days, the cold will destroy any lurking insects and their eggs. Follow the deep freeze with a thorough sifting, then repackage the grains in clear containers.

Insects and mold dote on whole grains, feasting well on their oils. Insects lay eggs on grain that are so tiny they can easily elude detection even when the grain is reduced to sifted flour. There may be times, however, when insects unmistakably proclaim their presence in your grain store. If you discover an infestation, either try the four-day freeze described above, or spread out the grain on a tray in a layer no deeper than ¾ inch thick, and heat it in a warm (140°F) oven for 30 minutes.

Grains are also prone to picking up the mold spores that inhabit our air. When the storage environment is cold and dry, however, mold doesn't have a chance. Although refrigerator storage is sufficient for grains you want to keep for only a few months, freezer storage is necessary if you want to store grains 18 to 24 months. Cooked grains can be frozen for 2 to 3 months, and cooked rice for 6 months.

While whole grains have a natural covering for their oils, whole grain flours are out there on the firing line in the battle against rancidity (yet another enemy of the much-beleaguered grains). With full exposure to the air on all sides, the oils in flour oxidize quickly and before you know it, the flour tastes rancid.

The fragility of whole grain flours may not be obvious to cooks used to highly refined, chemically treated flour. The bran and the germ that have been milled out of white flour are still present in whole grain flour. They are responsible for the superior nutritional quality, the nutty flavor and the vulnerability to warm temperatures of whole grain flours. Oats, cornmeal and buckwheat groats, while not flours, share the vulnerability of flour because they are so well endowed with oils that they, too, can become rancid very quickly.

Whole grain flours do not have quite the storage life of unmilled grains, but they will last twelve to eighteen months in the freezer. If you grind your own grains, it is of course best to do it just before you bake with them.

All grains are at their most nutritious when fresh. Freezing, however, does a very good job of helping them retain their nutritional quality. The protein and carbohydrate levels of grains, their main benefit, do not change no matter what the ambient storage temperature reads. The vitamin content, rather scanty to begin with in grains, is best retained at the low temperature and humidity level in the freezer.

The richer in oils the grain is to begin with, the more apt it is to lose its vitamin level at room temperature, and the better off it will be in the freezer. Even rice, a dry grain, retains a higher nutrient value when frozen than at room temperature. Moister grains like corn and wheat have substantially higher nutritional content when frozen.

You should freeze whole grains, flour and wheat germ in airtight, moisture proof freezer packaging, because all of them can easily absorb moisture and odors. Freeze flour in the packaging it comes in, overwrapped in a freezer bag. If you run short of freezer space, grains will transfer to the refrigerator more readily than other foods.

You can use uncooked grains straight from the freezer without thawing. Cooked grains are best thawed three to four hours in the refrigerator before re-heating over steam. Use flour without thawing, unless you are using it in combination with yeast. In that case, the dough will rise better if you first let the flour warm up to room temperature.

Pasta

Count cooked pasta among the good freezer foods. More than likely, you will store pasta as part of a cooked dish like lasagna. In that form, or within any other casserole, pasta moves from freezer to oven to table with ease.

Freezing pasta in prepared dishes is standard operating procedure. Cooking pasta simply to store it in the freezer on its own is hardly necessary; it takes so little time to cook it up fresh. If you have leftover cooked pasta and want to freeze

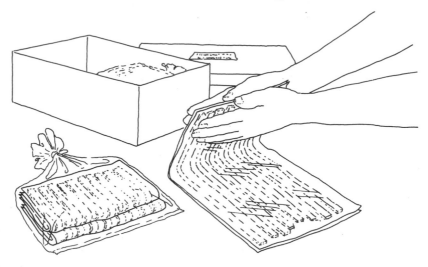

Lay the cut pieces of pasta on a sheet of wax paper and cover with another sheet. Fold the pile in half with the wax paper acting as a divider. Put the pasta in a plastic freezer bag and store the bags in a shoe box in the freezer to prevent breakage.

it, however, do so for two to three months, then reheat it by adding it directly to a pot of boiling water. As soon as the water returns to the boil, drain the pasta and serve it.

You should freeze your homemade pasta if you intend to store it for more than two days. Uncooked pasta freezes for twelve to eighteen months. Don't freeze the dough; freeze the shapes. Before putting the dough through the pasta machine for the last time, dust it with cornstarch or cornmeal, instead of wheat flour, and the surface won't be sticky. Lay the pasta out on wax paper. Cover it with another sheet of wax paper. Fold the noodles over so the top sheet of wax paper acts as a divider. Pack the noodles in a freezer bag. For extra protection, pack the freezer bag inside a shoe box. Let the pasta thaw at room temperature before cooking only until it is no longer brittle, about 15 minutes. Small pasta shapes can be used directly from the freezer.

Pasta frozen in small shapes will cook in about 4 minutes, 8 minutes if filled. Longer-shaped pieces, such as strips, will cook in about 5 to 6 minutes.

Nuts and Seeds

Nuts and seeds, small but powerfully good, owe their flavor to the richness of the oils they contain. Just as cooking oils, including those made from peanuts, walnuts, hazelnuts and sesame seeds, can become rancid, so can the nuts and seeds from which the oils are pressed. Light, heat and moisture activate the enzymes in nuts and seeds, making it inappropriate to keep them for a long time in open bowls or clear glass jars or on sunny kitchen countertops.

Use the Euell Gibbons method for cracking hickory nuts. Freeze them overnight, then crack each with a hammer along its edge. Swing hard enough to crack the shell but not so hard that you smash the nut.

Any nuts or seeds that you intend to keep for longer than a month belong in the freezer. You can freeze nuts chopped, ground, whole, roasted, spiced, in the shell or even in an unopened can. Nuts that are salted or spiced have a freezer storage life of six months, about half the time that plain nut meats can stay in the freezer. Seeds also freeze for six months. Toasting and roasting also diminish freezer storage life, and are best done when the nuts and seeds come out of the freezer, rather than before they go in, unless you intend to freeze them for only a few months. Unless freezer space is limited, it's best to freeze nuts in their shells, and crack them as you need them. Use nuts and seeds directly from the freezer, without thawing first.

Chestnuts

Chestnuts in their shells will last for a few months in the refrigerator, and for six months in the freezer. Once peeled, freeze them if you won't be using them within three days. To peel chestnuts easily, cut a slit in the flat side of each one, cover them with water and boil for about five minutes. Remove the chestnuts from the water one at a time, and cut off the shell while it is still damp.

Simmer the peeled chestnuts in stock, or in equal quantities of water and milk, until they're just tender. For compact freezer storage, puree them, while they're still warm, in a blender, food processor or food mill. The frozen puree will last for a year. Use it straight from the freezer or let it thaw and mix it together if it has separated at all. To prevent separation, add 1 tablespoon of butter or honey to each cup of puree before you freeze it.

To tempt your palate, combine chestnut puree and the chicken stock from your freezer to make good soup. Or remove chestnut puree and squash or pumpkin puree from your freezer, thaw and combine them to lend each food an interesting new dimension.

Soy Foods

Although soybeans are actually a legume, not a grain, the products made from them are distinctly unvegetable. We'll talk about freezing tofu, tempeh and other soy foods here.

Tofu

Tofu is the exception to so many other culinary rules, it might as well be the exception to some freezing rules, too. For one thing, tofu is a food that comes out of the freezer better than it went in. That's not supposed to happen. Foods are, at best, supposed to come out as good as they went in, but no better. Tofu goes into the freezer soft and cheeselike and it comes out firmer and drier. Its new consistency substantially increases the number of ways you can use it.

The firmer, chewier texture of frozen tofu makes it possible to skewer it. Its drier, spongier quality allows it to soak up sauces and marinades better. Frozen tofu will hold its shape in stews better than fresh tofu does. Or you can crumble it, season it, bind it with egg and use it as a base for stuffing. Or you can add that same mixture to a sauce and simmer it, and it will look and taste like ground meat.

Interestingly, tofu also changes color in the freezer, becoming a golden yellow. It fades to light beige as it thaws.

Tofu lasts in the refrigerator for ten days. If you intend to keep it any longer, or if you want to alter its texture, freeze it.

Rinse the tofu with water and pat it dry with a towel. Cut it into slices or cubes, depending on how you intend to use it later, or keep it in the original cake. Freeze it in a freezer bag with all the air squeezed out. In 48 hours, the texture transformation will have taken place. Tofu can stay in the freezer for up to six months. Thaw it overnight in the refrigerator, or in a colander at room temperature for two hours. Squeeze out the remaining water.

Tempeh

There are two points during the making of tempeh at which you may freeze it: after is is inoculated, and after it incubates. Freezing at either of these stages, however, will have some negative effects. If you freeze tempeh after it is inoculated, thaw it in the refrigerator and then set it out on the counter to come up to room temperature before incubating. Freezing will make the culture weaker. If you freeze tempeh after incubation, its texture will become a bit sticky and its flavor will become sharper. The stickiness will make it impossible to dry-fry the tempeh once it is thawed.

Other Soy Products

Soy flour has more oil than regular wheat flour and becomes rancid faster. Always keep it in the freezer during warm weather months, and use it directly from the freezer without thawing first. Whole, dry soybeans, like soy flour, belong in the freezer when the weather is warm and can be used directly from the freezer without thawing. Soy grits also freeze very well.

Soymilk can also be frozen, but it will curdle in the freezer unless you sweeten it first. Mix 2 tablespoons of honey or maple syrup into each quart of soymilk, then freeze.

SARA JO'S
BROCCOLI-RICE CASSEROLE

Freeze this casserole for 3 months if using fresh broccoli and rice. Do not freeze if using frozen broccoli or rice.

4 teaspoons butter	4 eggs
½ cup finely chopped onions	⅓ cup milk for frozen broccoli, ½ cup for fresh
3 cups chopped broccoli, partially thawed if frozen	dash of ground black pepper
1 clove garlic, crushed	freshly grated nutmeg, to taste
2 cups cooked brown rice, partially thawed if frozen	½ cup shredded mozzarella cheese
½ cup grated Parmesan cheese	

Heat butter in a large skillet. Add onions and sauté until golden but not brown, about 5 minutes. Add broccoli. Sauté 2 minutes. Add garlic, rice and Parmesan cheese.

Beat eggs slightly in a medium bowl. Add milk, pepper and nutmeg. Add this to the broccoli mixture. Turn into a buttered 2-quart ovenproof casserole. Casserole can be frozen at this point, or top with mozzarella cheese, cover and bake at 350°F for 20 to 25 minutes.

To make a frozen casserole, add cheese, cover and bake 30 to 45 minutes. Uncover and continue baking until cheese is golden brown and bubbly.

YIELD: 4 to 6 servings

BROWN RICE AND WALNUT STUFFING

A good stuffing for zucchini or eggplant. This stuffing can be frozen for 3 months if not made with frozen rice.

1 tablespoon vegetable oil	⅛ teaspoon cayenne pepper
1 tablespoon butter	1 cup chopped walnuts
1 large onion, chopped	⅓ cup wheat germ
1 cup brown rice, thawed if frozen	1 cup ricotta or cottage cheese
¼ cup minced fresh parsley	6 medium-size zucchini, or 3 medium-size eggplants
2 cups stock, thawed if frozen, or water	½ cup shredded mozzarella, Swiss or cheddar cheese, optional
2 tablespoons lemon juice	
1 teaspoon soy sauce	
1 teaspoon dried oregano	
½ teaspoon dried thyme, crushed	

In a large skillet, heat oil and butter and sauté onions until soft. Stir in rice, parsley, stock, lemon juice, soy sauce, oregano, thyme and cayenne. Bring to boil, cover, then simmer over low heat for 25 minutes. Stir in walnuts, wheat germ and ricotta or cottage cheese. At this point the stuffing may be cooled and frozen for later use.

Halve the zucchini or eggplants lengthwise. Steam zucchini for 8 to 10 minutes, eggplant for 15 to 20 minutes. Drain and cool, then scoop out flesh, leaving a ¼-inch shell.

Coarsely chop flesh and mix with rice mixture. Stuff the shells. Place in two shallow baking dishes. Add just enough water to cover the bottom of the dishes. Cover with foil. At this point, if the stuffing is fresh, the shells may be frozen for later use.

Bake, covered, at 350°F for 15 minutes or until the shells are tender, (about 30 to 45 minutes for frozen shells).

Uncover, sprinkle with cheese and leave in oven a few more minutes until cheese melts.

YIELD: 6 to 12 servings

BROWN RICE, WHEAT GERM AND GIBLET STUFFING

4 frozen chicken livers, or 1 turkey liver, thawed
2 tablespoons vegetable oil
½ cup chopped onions
½ cup chopped celery
3 cups cooked brown rice, thawed if frozen
½ cup wheat germ
½ cup chopped fresh parsley

1 teaspoon soy sauce
1 teaspoon dried sage
1 teaspoon dried thyme
⅛ teaspoon cayenne pepper
½–¾ cup frozen Chicken Stock (page 73) or turkey stock, thawed

Broil livers for 5 minutes, turning once, then chop coarsely. In a large skillet, heat oil and sauté onions and celery for 10 minutes.

In a large bowl, combine chopped liver, onion mixture, rice and wheat germ. Add the parsley, soy sauce, sage, thyme, cayenne and ½ cup stock. Add to skillet and heat through. If mixture appears too dry, add a little more stock. Use immediately.

YIELD: 5 cups

CRANBERRY-CHESTNUT STUFFING

The brisk tang of cranberries complements duck or goose as well as the more traditional turkey or chicken. Garnish with fresh mint leaves or grapes for a truly festive appearance. You can freeze this stuffing for 3 months if not made with frozen rice. Don't refreeze stuffing made with frozen rice.

3¾ cups cranberries	2 teaspoons dried oregano
boiling water	½ teaspoon dried rosemary
½ cup minced onions	pinch of cayenne
1 cup sliced celery	pepper
4 cloves garlic, minced	2½ cups cooked brown rice,
¼ cup sliced mushrooms	thawed if frozen
2 tablespoons vegetable oil	2 cups cooked and
2 tablespoons honey	peeled chestnuts
1 teaspoon dried sage	(page 291)
1 teaspoon ground coriander	

Cover cranberries with boiling water and let sit about 5 minutes. Drain and set aside.

In a medium skillet, sauté onions, celery, garlic and mushrooms in oil over medium heat until limp. In a large mixing bowl, combine sautéed vegetables, cranberries, honey, sage, coriander, oregano, rosemary, cayenne, rice and chestnuts. Use immediately, or, if using fresh rice, pack in containers and freeze.

Thaw stuffing completely in the refrigerator. Then immediately before roasting turkey or other fowl, stuff it loosely with the mixture. Or bake stuffing separately in an oiled baking dish at 350°F, covered, about 35 minutes.

YIELD: 10 cups

WHOLE WHEAT PASTA

This recipe will yield about 1 pound of fresh pasta that can be cut into any shape you like and frozen for 6 months.

1 cup whole wheat pastry flour	1 tablespoon olive oil
1 cup whole wheat flour	1–2 tablespoons warm water, optional
3 eggs	

Place flours in a mixing bowl, making a deep well in the center. Drop eggs into the well and beat them with a fork, slowly drawing some of the flour into the eggs. Add oil and continue gradually incorporating flour into mixture. Add water as necessary, mixing dough with hands until dough holds together but is not sticky.

Turn out dough onto a board. Knead dough for 5 minutes until smooth and elastic. Wrap in plastic wrap or wax paper and let rest for 1 hour.

If using a pasta machine, follow manufacturer's directions. Cut to 12-inch lengths, rolling dough as thin as possible.

For freezing, dust dough with cornstarch instead of wheat flour while rolling it out.

Cut pasta into desired shape and, without drying, place noodles in layers separated by wax paper. Fold if desired, place in freezer bags and freeze.

Thaw the pasta slightly before cooking; otherwise, it will be too brittle to work with. Fresh or slightly thawed, the pasta will cook in about 6 minutes.

YIELD: 48 ravioli, 4 to 8 servings of spaghetti

SPINACH LASAGNA

This lasagna goes right from the freezer to the oven without thawing. For a saucier lasagna, top with additional hot tomato sauce when serving. You can freeze this lasagna for up to 3 months.

3 cups Marinara Sauce (page 54) or Italian Tomato Sauce (page 56), thawed if frozen

4 cups shredded mozzarella cheese

1 pound ricotta cheese

¼ cup grated Parmesan cheese

¼ cup grated Romano cheese

2½ cups chopped cooked spinach, thawed if frozen, well drained

3 eggs, beaten

1 pound lasagna noodles, cooked

Spread 1½ cups sauce in a 9 × 13-inch shallow ovenproof casserole. In a large bowl, thoroughly mix the cheeses, spinach and eggs. Assemble layers in casserole, beginning with noodles, followed by cheese-spinach mixture, until all ingredients are used. Spread remaining 1½ cups of sauce over the top.

If freezing, cover with plastic wrap and aluminum foil and freeze immediately. Be sure to remove plastic wrap before baking.

Bake at 350°F for 45 to 50 minutes for unfrozen lasagna, or 1¾ to 2 hours, covered, if frozen.

YIELD: 8 to 10 servings

SPINACH PASTA

Can be frozen for 6 months.

½ cup chopped blanched spinach, thawed if frozen

4 teaspoons olive oil

2 eggs

1 cup whole wheat pastry flour

1 cup whole wheat flour

Place spinach and 1 teaspoon oil in a small saucepan and cook gently until liquid evaporates. Cool slightly. Puree in blender with eggs until smooth. Measure ¼ cup of the puree (there may be a bit leftover).

Place flours in a bowl, making a well in the center. Drop spinach mixture into the well and beat with a fork, slowly drawing some flour into the mixture. Add 3 teaspoons oil and continue gradually incorporating flour into mixture. Mix dough with hands until it forms a ball.

Knead dough on board for 5 minutes until smooth and elastic, using any leftover flour in bowl if necessary. Wrap in plastic wrap or wax paper and let rest for 1 hour.

Continue as directed for Whole Wheat Pasta (page 297).

YIELD: 48 ravioli, 4 to 8 servings spaghetti

GOLDEN STUFFED MANICOTTI

Cooked winter squash makes this manicotti a unique and tasty vegetarian main dish. Freeze it for 3 months if made with fresh squash. Do not freeze if made with frozen squash.

4 cups mashed cooked
     winter squash, thawed
     if frozen
	1 teaspoon freshly grated
	     nutmeg

1 cup grated Parmesan
     cheese
	12 whole wheat manicotti
	     shells, cooked

2 tablespoons minced fresh
     parsley
	2 cups Marinara Sauce
	     (page 54), thawed
	     if frozen

In a bowl, mix together squash, cheese, parsley and nutmeg. Using a spoon or a pastry bag fitted with a large tube, fill each manicotti shell with about 5 tablespoons of squash mixture.

Lightly butter a baking dish large enough to hold all the manicotti in one layer. Arrange shells in dish. Cover with sauce. Bake at 350°F for about 20 minutes, until heated through. Serve now or cool completely and freeze.

To make frozen manicotti, bake, covered, at 350°F for 40 minutes or until sauce is bubbly.

YIELD: 4 servings

CHEESE RAVIOLI

Freeze this ravioli for 4 months.

1 cup ricotta cheese
½ cup shredded mozzarella
    cheese
¾ cup grated Parmesan
    cheese
1 egg, beaten

1 teaspoon chopped fresh
    parsley
1 recipe Whole Wheat
    Pasta (page 297)
    or Spinach Pasta
    (page 298)

In a medium-size bowl, combine cheeses, egg and parsley. Mix thoroughly. Refrigerate until pasta is ready.

If using a pasta machine, roll dough out to less than ⅟₁₆ inch thick. Cut into 12-inch lengths. If rolling by hand, divide dough in half. Flour lightly and roll out both balls as thinly as possible.

Place rounded teaspoonfuls of filling about 1 inch apart on half of the dough. Moisten with water around edges and between mounds of filling. Place remaining dough on top of filling and lightly press dough together between the filling mounds.

Cut ravioli apart with a pastry cutter. Allow ravioli to air-dry for about 15 minutes. Leftover dough can be cut into other noodle shapes and frozen.

To serve immediately, cook in 4 quarts of boiling water for 3 to 5 minutes. Drain and serve with a tomato sauce or toss with butter and grated Parmesan cheese.

To freeze, place ravioli on baking sheets and freeze until firm. Layer ravioli between sheets of wax paper and place in freezer bag.

To cook after freezing, do not thaw; cook 8 minutes or until tender.

YIELD: 48 to 60 ravioli

BROCCOLI BAKE WITH PASTA

If made with fresh broccoli, this can be frozen for 3 months. Do not freeze if made with frozen broccoli.

2 cups chopped broccoli
1 pound whole wheat pasta shells, cooked
1 cup cottage or ricotta cheese
2 tablespoons wheat germ
¼ teaspoon ground black pepper

¼ teaspoon paprika
¼ teaspoon dry mustard
3 eggs, beaten
½ pound mozzarella cheese, shredded
¼ cup grated Parmesan cheese

Steam broccoli until crisp-tender, about 10 minutes for fresh, 5 minutes for frozen. Set aside.

Mix pasta, broccoli and cottage or ricotta cheese together in a large bowl. Set aside.

Butter an 11 × 7-inch baking dish. Sprinkle wheat germ over the bottom of the dish. In a medium bowl, beat together pepper, paprika and mustard with the eggs. Pour this mixture over pasta and combine well.

Spread pasta, broccoli and cheese mixture in the baking dish. Freeze now or top with mozzarella and Parmesan cheeses. Bake at 350°F for 30 minutes or until top is golden brown.

To reheat frozen casserole, add mozzarella and Parmesan cheeses, then cover and bake in a 350°F oven for 30 minutes. Uncover and bake for 15 minutes more.

YIELD: 4 servings

TUNA-STUFFED SHELLS
WITH SPINACH SAUCE

Can be frozen for 3 months.

1 cup cooked tuna, thawed if frozen	1 recipe Mochiko Rice Flour White Sauce, medium thickness (page 60), thawed if frozen
1½ cups ricotta cheese	
¼ cup grated Parmesan cheese	
1 egg, lightly beaten	2 cups chopped spinach, thawed if frozen, drained well
1½ teaspoons dried tarragon	
24 large whole wheat pasta shells, cooked	
1 tablespoon butter	1 teaspoon vegetable seasoning
1 large onion, chopped	1 teaspoon soy sauce
1 clove garlic, minced	1½ cups shredded Gruyère or Jarlsberg cheese

In a medium-size bowl, combine tuna, ricotta and Parmesan cheeses, egg and ½ teaspoon tarragon. Mix to combine. Spoon this mixture into the shells. Cover with damp towel.

In a medium skillet, melt butter; add onions and garlic and sauté until limp. Set aside.

Prepare white sauce according to directions. Stir in remaining tarragon, spinach, onions and garlic, vegetable seasoning and soy sauce. Puree in food processor or blender until smooth.

Pour one-third of the spinach sauce into a 9 × 13-inch shallow ovenproof casserole. Arrange stuffed shells on top. Pour remaining sauce over all. Freeze now, or top with Gruyère cheese, cover with foil and bake in 350°F oven 30 to 35 minutes.

To make frozen shells, top with Gruyère cheese, then cover and bake for 40 minutes at 350°F, uncover and bake 15 minutes more.

YIELD: 4 servings

VARIATION: Substitute dillweed for the tarragon.

CHAPTER 12

BAKED GOODS

The baked goods we relish are the warm, freshly made breads, the cookies snatched right off the cookie sheet, the steaming pies straight out of the oven. To hold onto that just-made quality in baked goods, it is better to freeze them after they are baked, rather than before. Baked goods freeze without noticeable change in taste or texture. However, with the exception of pie crusts and cookie dough, texture suffers when you freeze unbaked pastry and dough. It is possible to freeze unbaked dough and pastry, but you have to accept that the finished product will not be quite the same as freshly baked goods.

Freezing Bread and Rolls

All baked breads and rolls, no matter what their size or shape, freeze beautifully. Allow them to cool completely on a rack, then wrap and freeze them immediately. If you wrap them in foil they can go straight from the freezer to a warm oven or the toaster. You can, of course, slice and butter or season your bread and rolls before freezing them, then reheat and serve. Spread bread with butter and garlic or grated cheese, and freeze it as a ready accompaniment to an evening's dinner. Or keep in the freezer bags of bread that consist of slices cut from several different loaves. Reheated in the oven, they'll provide a basketful of

The refrigerator is no place for bread. It quickly goes stale there from evaporation loss. For the short term, keep bread, wrapped, at room temperature. For the long term, freeze it.

selections on the table. Cutting bread and rolls is easier when they are still partially frozen. This is particularly true of crumbly items like fruit and nut breads.

Breads and rolls have the longest storage life of any baked goods, but they're best consumed within several months of being frozen because their flavor begins to deteriorate.

You can buy and freeze day-old bread and it will come out of the freezer as if it were fresh. Freeze the special regional breads you buy when you travel, too. And freeze crumbs made from day-old bread to use for toppings and fillings.

Freezing Yeast

Compressed yeast is active and spoils quickly. Refrigerate it if you will use it within three weeks, otherwise freeze it, and it will last for two months. When you want to bake with it, thaw out as much as you need overnight in the refrigerator.

Active dry yeast in moistureproof packages keeps for several months and does not need freezing, though refrigerating it prolongs its life.

Freezing Dough for Breads and Rolls

The yeast in batter slowly dies in the freezer. If you freeze your bread dough and use it within ten days (or seven days in the case of rolls), the results will not be substantially different because not enough of the yeast would have died. When you bake the loaf, you can expect that the dough will not rise as much as you're used to and the resulting texture will be heavier. Recipes created specifically for frozen dough compensate for the yeast being gradually killed by calling for more yeast. For example, the recipe for Freezer Sandwich Bread on page 308 allows you to keep the dough in the freezer for up to four weeks.

To freeze bread dough, prepare the dough as you normally would to the point where the loaves are formed. Line the bread pans with aluminum foil. Shape the loaves loosely in the pans. Make the loaves 2 inches thick at most, so that they will thaw that much more rapidly. Cover the pans with plastic wrap and freeze them. When the loaves are frozen, remove them from the pans, wrap each loaf in foil or plastic wrap and return it to the freezer. Remove the dough within four weeks, place it in a greased baking pan, cover it with a towel and let it thaw and double in volume for about six or seven hours. Bake as usual. You can expect bread made this way to dry out very quickly; eat it soon.

To freeze the dough for rolls, either freeze them on a tray or in muffin tins, covered with plastic wrap. When they are frozen, after about four hours wrap the rolls individually and freeze them. Remove the rolls within a week, cover them with a cloth, let them double in volume in a warm place for about 2½ hours, then bake as usual.

Biscuits made from frozen dough may be smaller and tougher than you would like. But if you do want to freeze biscuit dough, handle it in the same way as the dough for rolls. Let the biscuits thaw for an hour, wrapped, at room temperature, then bake them as you normally would. It's really far better to freeze your baked biscuits, then reheat them straight from the freezer in a 325°F oven for 15 to 25 minutes.

Quick breads contain no yeast; therefore, they require no risings, and are quick to make, to bake and to be eaten. Thaw coffee cakes and fruit and nut quick breads at room temperature or, wrapped in foil, in a 325°F oven for 15 to 25 minutes. If you freeze batter for fruit and nut quick breads, thaw it at room temperature, then bake immediately, according to the recipe.

Freezing Cakes

It's best to bake cakes (or muffins or cupcakes), freeze them, thaw them, then fill and frost them. Any other sequence will have drawbacks. There's no particular advantage to freezing batter, and several disadvantages. For one thing, the batter will very likely lack volume and quality when it is baked. Besides, it takes less time to prepare batter fresh than it does to thaw it out.

It's best to freeze cake fillings and frostings separately from the cakes themselves. Fillings makes cakes get soggy in the freezer. Frosting may crack in the freezer, and will certainly shorten the cake's storage time and elongate its thawing time. An unfrosted cake, on the other hand, freezes well for six months and thaws at room temperature in an hour. A frosted cake will keep for four months and takes one to two hours to thaw at room temperature. Besides, cakes look fresher if you frost them just before serving.

Freeze unfrosted cakes until they are firm enough to cut without crumbling. Create shapes to honor a special occasion. Frost the shapes and serve them at the celebration.

If you do decide to freeze a frosted cake, even after all these discouraging facts (or because you have leftovers), put it unwrapped in the freezer to firm the icing. Then, put wax paper over the iced parts and wrap the whole cake in freezer paper. To protect it further, put it in a box. Thaw it unwrapped in a covered cake dish.

An unfrosted cake, wrapped in foil, can go straight from the freezer to a 250° to 300°F oven for 20 to 25 minutes. The gentle heat will redistribute the moisture in the cake and give it a fresh taste. You can also thaw cake successfully in the refrigerator in three to four hours.

Muffins and cupcakes are best baked and then frozen for up to three months. They'll thaw at room temperature in 30 minutes, or you can reheat them in a 300°F oven for 20 minutes.

Freezing Cookies

Freeze your cookies, but bear in mind that some will break if they're crisp, no matter how well you wrap them. Pack them in a cookie tin, or a box, with foil or plastic wrap between the layers, and freeze them for up to four months. They'll thaw at room temperature in ten minutes if you unpack them, and you'll never know they'd been frozen. Reheat them quickly in a 350°F oven.

Freezing cookie dough is an economical use of space, and is generally an exception to the rule that baked goods should be frozen after baking rather than before. Unbaked cookies can be frozen up to four months. If you make refrigerator cookies, you can cut out the cookies, tray-freeze them, then repackage them for the freezer with the layers of cookies separated by freezer wrap. Don't pack so many layers that you crush the dough.

Another method is to freeze cookie dough in the shape of a long roll. Thaw it just enough to cut out cookies from it when you are ready to bake. Or you can pack the dough into empty juice cans, with both ends removed. To use the dough, push it out and slice it. Then bake the cookies as you normally would. Most cookies bake in about 12 minutes in a 350°F oven.

Drop the dough for drop cookies from a spoon straight onto a lightly greased cookie sheet and freeze. Repackage the frozen, unbaked cookies in freezer containers. Bake the cookies without thawing.

Freezing Waffles, Pancakes and Crepes

There's no point in freezing waffle and pancake batter, unless it is leftover. It's easier to freeze waffles and pancakes after they're made and slip them in the toaster or 400°F oven for two to three minutes to reheat straight from the freezer.

Crepe batter will freeze, too, but so will the crepes themselves. All of the above should be frozen with layers of freezer wrap in between them to act as dividers. All should be tightly overwrapped. They can be frozen three to four months.

Freezing Pies and Pastry

Pies and tarts made with a lot of fat or shortening freeze best, either baked or unbaked. To freeze pie and tart dough unbaked, prepare it and line a pan with it. Freeze it in the pan, then remove and repackage it, ideally stacked with other pie shells in a sturdy box. It will freeze for two months. Bake straight from the freezer at 425°F for 12 to 15 minutes, then cool and fill. Or defrost at room temperature for 20 minutes, fill and bake as usual.

You can also freeze pie dough rolled out and ready to go, but not molded into a pie shell form. Stack several circles this way, with freezer wrap separating each one. They will thaw in 15 minutes. Or, shape the dough into a ball, wrap it in plastic wrap and freeze it for up to four months. Thaw at room temperature three to four hours, sprinkle it lightly with flour and roll it out.

Use your freezer to firm up too-sticky dough. Scoop the dough onto wax paper. Cover it and press it into a circle less than an inch thick. Freeze for 20 to 30 minutes, then roll it out with ease.

You can also freeze a pie with a fruit filling in an unbaked crust for two months. Brush the bottom crust with egg white before you add the filling. The egg white will act as a barrier to keep the filling from making the crust soggy. Do the same for the top crust on the side that will press against the fruit. Don't glaze the top crust with milk, water or egg. Cut slits in the upper crust of the pie just before you place it, unthawed, in the oven. Baking time varies with the type of pie filling, of course, but as a general guideline, bake a frozen pie at 450°F for 15 to 20 minutes, then at 375°F for about 45 minutes to one hour.

Freezing Baked Pastry Dough

To bake a pie shell before freezing, fit the dough gently into the pie pan. Prick the bottom and sides with a fork. Freeze it for 15 minutes if you can, to help prevent shrinkage. Bake the pie shell at 425°F for 12 to 15 minutes, turning it so it will bake and brown evenly. Baked pie crusts will freeze for four months and thaw at room temperature in 20 minutes.

Freeze the cooled pie crust, remove it from the pan and wrap, or freeze it in a foil pan. To form a cover, invert another pie pan over the pie and tape the edges securely all around with freezer tape.

You can also bake filled pies before you freeze them. Fruit and mincemeat pies freeze best. Pumpkin and squash pies do well, too. But don't freeze cream or custard pies. Most pies can be baked in a 425°F oven for 45 to 60 minutes, turning occasionally so the top crust and rim brown evenly. Then cool, wrap and freeze the pie for up to four months.

Reheat frozen pies at 400°F for 30 to 40 minutes. Or thaw at room temperature for two or three hours and serve.

Which way is better, freezing pies baked or unbaked? It's anyone's guess. Frozen, unbaked pies have a very fresh taste when baked. The shells are good and flaky. Baked pies store for twice as long as unbaked pies. Both are highly satisfactory storage methods.

FREEZER SANDWICH BREAD

You can freeze this dough for 4 weeks or the baked bread for 3 months.

¼ cup butter	2 tablespoons lemon juice
½ cup finely chopped onions	2 tablespoons active dry yeast
2 cups warm water	6-6½ cups whole wheat flour
⅓ cup plus ½ teaspoon honey	

Melt butter in a small skillet over medium heat. Add onions and sauté for 5 minutes, or until soft. Remove from heat and set aside to cool.

Combine 1½ cups of the warm water, ⅓ cup of the honey and lemon juice in a large bowl.

In a small bowl, mix remaining ½ teaspoon honey into remaining ½ cup warm water. Add yeast and stir to dissolve. Allow to stand for 2 to 3 minutes to proof. It will become foamy.

Add 2 cups of the flour, onion mixture and yeast mixture to the water-lemon juice mixture in the bowl. Stir until smooth. Add enough of the remaining flour a little at a time to make a dough that is easy to handle and not sticky. Place dough on a lightly floured surface and knead for 10 minutes.

Place dough in an oiled bowl, turn dough to put oiled side up, cover and put bowl in a warm spot for about 30 minutes. Punch down dough, divide in half and form each piece into a loaf. The dough can be frozen at this point in plastic freezer bags.

To bake immediately, place loaves in oiled 9 × 5-inch loaf pans and allow to rise until almost doubled in bulk, about 1 hour. Bake in a 350°F oven for 35 to 45 minutes, or until bread tests done. Cool on wire racks.

To use frozen dough, remove from freezer bags and place in the oiled pans. Allow to thaw and rise, about 4 hours. Bake as directed.

YIELD: two 9 × 5-inch loaves

WHOLE WHEAT PEAR BREAD

Can be frozen for 3 months.

1 cup Pear Sauce (page 280), thawed if frozen	2 cups whole wheat pastry flour
2 tablespoons vegetable oil	1 teaspoon baking powder
1 teaspoon grated lemon peel	1 teaspoon baking soda
½ cup honey	½ teaspoon ground cinnamon
2 tablespoons lemon juice	¼ teaspoon ground cloves
⅓ cup water	1 cup chopped almonds
1 egg, beaten	

Combine pear sauce, oil, lemon peel and honey in a large bowl. Add lemon juice, water and egg and mix well.

Mix together flour, baking powder, baking soda, cinnamon and cloves.

Add flour mixture to liquid mixture and stir just enough to moisten flour. Stir in almonds.

Pour batter into a buttered 9 × 5-inch loaf pan. Bake at 325°F for 60 to 65 minutes.

Cool bread 10 minutes in pan. Remove from pan and cool on wire rack completely before slicing and serving or freezing.

Thaw frozen bread at room temperature for 3 to 4 hours.

YIELD: one 9 × 5-inch loaf

WHEAT AND CHEESE CRESCENT ROLLS

These easy, no-knead dinner rolls are good for a first-time baker. They can be frozen for 3 months.

4 teaspoons active dry yeast	¼ cup water
½ cup lukewarm water	3 eggs
3 tablespoons plus 1 teaspoon honey	2 cups whole wheat flour
1 cup vegetable oil	2 cups whole wheat pastry flour
1 cup warm buttermilk	2 cups grated Parmesan cheese

In a small bowl, dissolve yeast in the lukewarm water. Add 1 teaspoon honey. Allow the yeast mixture to proof, about 10 minutes.

In a large bowl, mix oil, 3 tablespoons honey, buttermilk and water. Beat 2 eggs and add to oil mixture. Add dissolved yeast mixture. Gradually stir in flours and cheese, mixing well. At this point, dough will be very soft. Cover and place in refrigerator to chill until firm, about 1 hour.

Divide dough into three equal parts and roll each one out on a floured surface into a large circle ⅛ inch thick. Beat remaining egg and lightly brush dough.

Cut each circle into twelve wedges using a pizza cutter, pastry wheel or sharp knife. Roll each wedge toward the center. Place point side down to form in a crescent shape. Brush tops lightly with beaten egg.

Place on buttered baking sheet, leaving enough room for each crescent to rise. Let rise in a draft-free place for 1½ hours.

Bake at 400°F for 12 to 15 minutes, until golden brown. Serve warm, or cool completely and freeze.

To reheat frozen rolls, wrap in foil and place in 400°F oven for 5 to 7 minutes.

YIELD: 36 rolls

WHOLE WHEAT PITA BREAD

Pitas are great for making pocket sandwiches. They can also be broken into large pieces and used as scoops for dips. Freeze pitas for 4 months.

2 teaspoons active dry yeast	3 teaspoons honey
1¼ cups lukewarm water	3-3½ cups whole wheat flour
	1½ teaspoons vegetable oil

Combine yeast, water and 1 teaspoon of the honey in a large bowl. Let stand 5 minutes. Stir in remaining honey and 3 cups flour.

Knead dough until smooth, about 10 minutes. Add remaining flour if dough is too sticky.

Coat the bowl with oil. Place dough in bowl and turn so the oiled side is up. Cover bowl with a damp towel and set in a warm place until doubled in bulk, about 1 hour.

Punch down dough and knead briefly. Divide into twelve pieces and roll each into a ball. Place on a lightly greased baking sheet and cover with a damp towel. Set in a warm place for 15 minutes.

Preheat oven to 500°F.

On a lightly floured surface, roll each ball into a circle ⅛ inch thick, about 5½ inches in diameter.

Preheat a baking sheet in the oven for 5 minutes.

Carefully place two or three pitas on the sheet and place in oven. Close the door immediately.

Bake until pitas are puffed and light brown, about 5 minutes.

Remove pitas from baking sheet, wrap in a towel and place in paper bag for 15 minutes. This prevents them from deflating. Repeat with remaining pitas.

Use immediately or cool and store in refrigerator or freezer.

To reheat, thaw completely if frozen and use as is, or toast in toaster or heat in 350°F oven for 10 minutes.

YIELD: 12 pitas

NOTES: For a fast pizza, split pitas in half. Cover with pizza sauce, cheese and a favorite topping. Bake until cheese is melted.

To make an easy appetizer, split pitas and spread with garlic butter. Cut into wedges and sprinkle with shredded cheese. Bake until lightly browned.

VARIATION: Separate the dough into six pieces instead of twelve.

HAMBURGER ROLLS

A freezer full of rolls can provide the basis for many fast meals. For instance, a split roll, filled with cooked meats, vegetables and tomatoes, can be topped with cheese and baked until hot and bubbly. This recipe is easy to make and can be adjusted to make hot dog and steak rolls. You can freeze the rolls for 6 months.

1¾ cups warm water	2 eggs, lightly beaten
½ cup vegetable oil	6-6⅓ cups whole wheat
⅓ cup honey	flour
3 tablespoons active dry yeast	

In a large bowl, mix together water, oil, honey and yeast. Let stand 15 minutes. Stir in eggs and 6 cups flour. Knead about 7 minutes until smooth. If dough is sticky, add remaining ⅓ cup flour a little at a time till the dough is smooth.

Roll out on floured surface until ¾ inch thick. Cut with a 4-inch round cutter.

Place rolls on oiled baking sheet. Cover with towel and let rise 10 minutes.

Bake at 425°F for 8 minutes. Cover the partially baked rolls with foil so rolls do not over-brown. Bake an additional 7 minutes.

Remove rolls immediately from pan and place on racks to cool. Cool completely before placing in freezer bags.

Thaw frozen rolls at room temperature for 2 hours before use.

YIELD: 12 rolls

VARIATIONS: To make hot dog rolls, follow the recipe above except roll dough into a rectangle 5 inches wide, 4 inches long and 1½ inches thick. Cut dough into twelve slices across the width, place on side and set to rise for 10 minutes. Bake as for hamburger rolls. Makes twelve rolls.

To make steak rolls, follow the recipe above except roll dough into a rectangle 8 inches wide, 10 inches long and 2 inches thick. Cut dough into eight slices, place on side on baking sheet and continue as directed for hamburger rolls. Makes eight rolls.

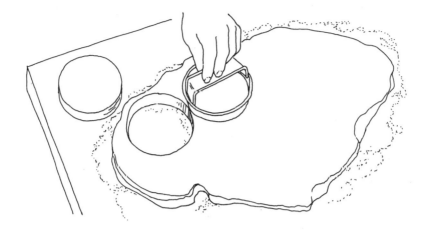

To cut dough for hamburger rolls, roll out the dough on a floured baking board to about ¾-inch thick. Cut rounds with a 4-inch cutter (or use a glass or cup). Transfer to an oiled baking sheet.

HERBED CRUMB TOPPING

Use these crumbs to dress up frozen vegetables or au gratin dishes. The crumbs can be prepared and frozen and will stay loose to use as needed. They can be frozen for 4 months.

2 slices lightly toasted bread
⅓ cup grated Parmesan or Romano cheese
½ cup grated mild cheddar cheese

¼ teaspoon onion powder
¼ teaspoon garlic powder
1 teaspoon dried basil
¼ cup vegetable oil

Place bread in blender or food processor and grind into coarse bread crumbs. Pour into a small bowl.

Mix cheeses, onion powder, garlic powder, basil and oil thoroughly with the bread crumbs. Freeze.

To use frozen crumbs, just sprinkle them on top of steamed vegetables and broil until crumbs brown. Or, sprinkle on top of an au gratin dish during the last 10 minutes of baking.

YIELD: 2 cups

CRUMB TOPPINGS

Crumb toppings are a "quick fix" to top fruit pies, or for sliced fruit baked as a fruit crisp. They store well in the freezer for 3 months. These recipes make 3 cups, enough for two to three pies.

In addition to crumb toppings, the same crumbs can be moistened with additional melted butter, about 2 tablespoons butter to 1 cup of crumbs, and used to make crumb crusts. Press buttered crumbs into pie pan, then chill or freeze, or bake until lightly browned, about 10 minutes at 350°F.

Thaw baked frozen crust and fill. For an unbaked frozen shell, bake frozen in a 350° oven for 15 to 20 minutes.

Coconut Fruit Topping

2 cups cornflakes, coarsely crumbled
⅓ cup toasted wheat germ
⅓ cup shredded coconut
⅓ cup diced dried apricots or peaches
1 tablespoon vegetable oil
1 tablespoon honey

Mix all ingredients well.

YIELD: about 3 cups

Maple Nut Topping

2 cups toasted whole wheat bread crumbs
1 cup finely chopped pecans
¼ cup melted butter
2 tablespoons maple syrup

Mix all ingredients thoroughly.

YIELD: about 3 cups

Nut and Seed Topping

2 cups rolled oats
½ cup finely sliced raw almonds
½ cup raw hulled sunflower seeds
⅓ cup melted butter
½ teaspoon ground cinnamon

Grind rolled oats in a blender or food processor until medium-fine texture. Combine all ingredients and bake in a shallow pan in 325°F oven for 20 to 30 minutes, until slightly browned (it should be slightly underbaked).

YIELD: about 3 cups

RAISIN-SPICE SQUASH BREAD

This hearty bread, similar to fruitcake, makes an excellent holiday gift. The batter can also be used to make muffins. The bread can be frozen for 3 months.

2½ cups whole wheat pastry
    flour
2½ teaspoons baking powder
¼ teaspoon ground cloves
¼ teaspoon ground nutmeg
1 teaspoon ground
    cinnamon
½ teaspoon ground ginger
¼ cup butter, at room
    temperature

½ cup light molasses
2 eggs
2 cups mashed cooked
    winter squash, thawed
    if frozen
1 cup golden raisins
¾ cup chopped hazelnuts

Sift flour, baking powder, cloves, nutmeg, cinnamon and ginger into a medium-size bowl.

Cream butter in a large bowl. Slowly beat in molasses and eggs, one at a time (the batter will separate a bit). Add squash and mix well.

Preheat oven to 350°F.

Add the flour mixture, ½ cup at a time, to the squash mixture, stirring only enough to combine. Fold in raisins and hazelnuts. Turn into a buttered 9 × 5-inch loaf pan and bake 50 to 65 minutes, until a cake tester comes out clean. Cool 5 minutes. Remove from pan. Cool to room temperature on a wire rack. Serve immediately or wrap and freeze.

Frozen bread can be thawed at room temperature for 3 to 4 hours.

YIELD: one 9 × 5-inch loaf

BANANA BREAD

Moist, delicious and very high in protein, this bread can be frozen for 3 months.

3 tablespoons butter, at room temperature	1 teaspoon vanilla extract
⅓ cup honey	¼ cup powdered milk
3 eggs, beaten	2 teaspoons baking powder
1 cup mashed banana, thawed if frozen	1 teaspoon baking soda
½ cup orange juice or water	2 cups whole wheat pastry flour
	½ cup chopped walnuts
	½ cup raisins

In a large bowl, beat butter and honey together until light. Add eggs, banana, orange juice and vanilla. Beat well.

In a medium-size bowl, combine powdered milk, baking powder, baking soda and flour. Stir to mix together. Add the dry mixture to the banana mixture, blending with a few strokes. Add walnuts and raisins; blend again, lightly.

Turn batter into a buttered 9 × 5-inch loaf pan. Bake at 325°F for about 1 hour, or until well browned and a cake tester comes out clean. Cool 10 minutes before removing from pan. Cool completely before slicing or freezing.

This bread can be thawed at room temperature for 3 to 4 hours.

YIELD: one 9 × 5-inch loaf

APPLESAUCE MUFFINS

You can freeze these muffins for 3 months.

1¾ cups whole wheat pastry flour	1 egg, beaten
2 teaspoons baking powder	½ cup milk
¾ teaspoon ground cinnamon	⅓ cup vegetable oil or melted butter
¼ teaspoon ground ginger	¾ cup Applesauce (page 263), thawed if frozen
¼ cup honey	

In a large bowl, combine flour, baking powder, cinnamon and ginger.

In another bowl, combine honey, egg, milk, oil and applesauce.

Add applesauce mixture to the dry ingredients, stirring until the dry ingredients are barely moistened. Do not overmix. The batter should be lumpy.

Fill buttered muffin tins two-thirds full. Bake at 400°F for 20 to 25 minutes, until lightly browned. Serve hot or cold or wrap and freeze.

Thaw frozen muffins at room temperature for 1 to 2 hours.

YIELD: 12 large muffins

BLUEBERRY-APPLE COFFEE CAKE

You can freeze this coffee cake up to 3 months.

⅔ cup honey
2 tablespoons frozen apple
    juice concentrate
½ cup vegetable oil
3 eggs
1 teaspoon vanilla extract
2 cups whole wheat pastry
    flour
1 teaspoon baking soda
1 teaspoon ground
    cinnamon

½ teaspoon ground nutmeg
2 cups sliced cooking
    apples, thawed if
    frozen
1⅓ cups blueberries, thawed
    if frozen
½ cup coarsely chopped
    walnuts

Using an electric mixer, whip honey at high speed, 3 to 5 minutes, until white and opaque. Add juice concentrate and oil and beat well. Add eggs and vanilla and beat well.

Sift flour, baking soda, cinnamon and nutmeg into a medium-size bowl. Add to the honey mixture and stir until thoroughly combined. Stir in apples, blueberries and nuts.

Pour into a buttered 9 × 13-inch baking pan. Bake at 350°F until lightly browned, 30 to 35 minutes. Cool in pan and serve, or freeze for serving later.

Thaw this cake at room temperature for 3 to 4 hours. If desired, the cake can be warmed, covered, in a 350°F oven for about 30 minutes.

YIELD: one 9 × 13-inch cake

BUTTERNUT SQUASH CHEESECAKE
WITH PECAN SHORTBREAD CRUST

Although butternut squash is especially good, acorn squash also works well in the cream cheese filling. Serve this rich dessert at Thanksgiving in place of pumpkin, pecan or mincemeat pie. You can freeze the assembled crust dough for 3 months, the filling for 1 month. If the pie is made with fresh crust and filling, you can freeze the baked pie for 1 month.

Crust

¼ cup cold butter	1 tablespoon honey
⅓ cup whole wheat pastry flour	½ teaspoon vanilla extract
1½ cups finely chopped pecans	

Filling

8 ounces cream cheese, at room temperature	½ cup maple syrup
2 eggs	2 teaspoons vanilla extract
1 egg yolk	¼ teaspoon almond extract
1½ cups butternut squash puree, thawed if frozen	

½ cup sour cream

Prepare crust by cutting the butter into the flour with a pastry blender until the mixture is in small crumbs. Stir in the pecans. Drizzle with the honey and vanilla. Stir to form a soft dough.

Freeze now for later use or pat dough evenly over the bottom of 9-inch springform pan. If dough is sticky, sprinkle lightly with flour. Refrigerate, covered, while preparing filling.

To make the filling, in a medium-size bowl, beat cream cheese until light and fluffy. Beat in, one at a time, the eggs and egg yolk. Stir in the squash puree, maple syrup and vanilla and almond extracts. Freeze filling now if desired.

Bake pecan crust in a 400°F oven until it begins to brown, 10 to 12 minutes. Pour in squash mixture, bake 10 minutes, then reduce heat to

325°F. Bake until center of cake is barely set, about 30 minutes. Turn off oven, open door slightly and let cheesecake cool in oven for 1 hour; then cool to room temperature on a counter.

Spread sour cream over cake. Chill thoroughly before cutting. The entire cake can be frozen at this point if all fresh ingredients were used.

To use frozen pastry, thaw for 6 hours at room temperature, or overnight in the refrigerator, then proceed as above. Thaw filling completely, then pour into pie shell and bake as directed.

Thaw an assembled frozen pie in the refrigerator overnight.

YIELD: one 9-inch cheesecake

VARIATION: Walnuts can be used in place of pecans in the crust.

BASIC LAYER CAKE

Can be frozen for 3 months.

2¾ cups whole wheat pastry flour	¾ cup honey
1½ teaspoons baking soda	1 cup sour cream
⅔ cup vegetable oil	1 teaspoon vanilla extract
	3 eggs

In a large bowl, mix together flour and baking soda. At medium speed, beat in oil, honey, sour cream, vanilla and eggs until thoroughly mixed.

Pour batter into two buttered and floured 9-inch round cake pans, and bake in a 375°F oven for 18 to 20 minutes or until toothpick or cake tester inserted in center of cake comes out clean.

Remove from oven, cool on wire racks for 5 minutes, then flip cakes out of the pans onto wire racks.

Cool completely. Decorate as desired or wrap in freezer paper or place in freezer bags.

Thaw the frozen layers at room temperature for 1 to 1½ hours, then frost.

YIELD: two 9-inch layers

ALMOND-RAISIN TORTE

You can freeze this cake unfilled for up to 3 months. The filling can be frozen separately up to 9 months.

Torte

9 egg yolks	¾ cup whole wheat bread
¾ cup honey	crumbs
2 tablespoons grated	¾ teaspoon almond extract
orange peel	1 cup egg whites (about 8
½ cup orange juice	egg whites), thawed
1½ teaspoons ground	if frozen
cinnamon	
1½ cups ground unblanched	
almonds	

Filling

¼ cup honey	½ cup water
1 tablespoon grated	1½ cups raisins
orange peel	¾ cup slivered blanched
½ cup orange juice	almonds, toasted
2 tablespoons Mochiko	(page 54)
rice flour	½ teaspoon vanilla extract

To make the torte, in a large bowl, beat egg yolks, gradually adding honey. Beat until creamy. Add orange peel and juice, cinnamon, almonds, bread crumbs and almond extract.

In a separate glass or metal bowl, beat the egg whites until they are stiff. Gently fold into egg yolk batter until thoroughly blended.

Pour batter into three 8-inch round cake pans, bottoms buttered and lined with wax paper. Do not butter pan sides. Bake at 350°F, 20 to 25 minutes, until top is golden brown. Cool completely and remove from pans. The cakes can be frozen at this point.

To prepare the filling, in a small saucepan, combine honey, orange peel and juice, rice flour, water and raisins. Simmer for 5 minutes over medium heat, stirring constantly, until mixture is thickened and easily stirred. Remove from heat, stir in almonds and vanilla.

Cool and freeze at this point if you wish.

Thaw cakes at room temperature for 1 to 2 hours, then proceed with filling.

To use frozen filling, thaw to room temperature. Divide into thirds before spreading on thawed cake layers.

YIELD: one three-layer 8-inch cake

PECAN CHIFFON CAKE

This light cake is a good way to use extra frozen egg whites and egg yolks. It can be frozen for 2 months.

2¼ cups whole wheat pastry flour	2 teaspoons vanilla extract
1 tablespoon baking powder	¾ cup cold water
1 teaspoon ground cinnamon	1 cup egg whites (about 8 egg whites), thawed if frozen
½ cup vegetable oil	½ teaspoon cream of tartar
5 egg yolks, thawed if frozen	1 cup finely chopped pecans
⅔ cup maple syrup	

Sift together flour, baking powder and cinnamon in a medium-size bowl.

In a large bowl, use an electric mixer to beat oil, egg yolks, maple syrup, vanilla and water until smooth. Slowly blend flour mixture into oil mixture.

Place egg whites in a large bowl, sprinkle with the cream of tartar, and beat until stiff.

Pour flour-oil mixture gradually over beaten egg whites and fold with rubber spatula until blended. Fold in pecans.

Pour batter into ungreased 10-inch tube pan. Bake at 325°F until top springs back lightly, about 65 minutes. Invert pan onto funnel or bottle until cake is cool. Remove from pan. Serve immediately or wrap and freeze.

This cake can be thawed at room temperature for 3 to 4 hours.

YIELD: one 10-inch cake

PUMPKIN ROLL

This mouth-watering dessert can be frozen for 2 months.

Cake

3 eggs	1 teaspoon baking powder
½ cup honey	2 teaspoons ground
¾ cup mashed cooked	cinnamon
pumpkin, thawed if	1 teaspoon ground ginger
frozen	½ teaspoon ground nutmeg
1 teaspoon lemon juice	½ cup finely chopped nuts,
¾ cup whole wheat pastry	optional
flour	

Filling

6 ounces cream cheese, at	½ teaspoon vanilla extract
room temperature	½ teaspoon lemon juice
¼ cup honey	
2 tablespoons chopped	
nuts	

To make the cake, beat eggs in a large bowl with an electric mixer on high speed for 5 minutes, until very light and fluffy. Beat in honey, pumpkin and lemon juice.

In another small bowl, combine flour, baking powder, cinnamon, ginger and nutmeg. Fold into pumpkin mixture, stirring just until ingredients are combined. Do not overmix.

Butter a 10 × 15-inch jellyroll pan. Line with wax paper or parchment. Butter the paper. Turn pumpkin mixture into pan, smoothing with a spatula. Sprinkle with nuts, if desired. Bake at 350°F for 15 minutes.

Invert cake onto a tea towel. Peel paper off bottom of cake. Immediately roll up cake, using tea towel underneath to help roll it. Set aside to cool.

Make the filling by beating together cream cheese, honey, vanilla and lemon juice.

When cake is cool, unroll it and spread with about three-quarters of the filling. Reroll cake. Spread remaining filling over top. Sprinkle with nuts.

Finished roll can be frozen. Allow to stand at room temperature 30 minutes before serving.

YIELD: 8 servings

MOLASSES COOKIES

These traditional Christmas cookies may be made with or without anise, to suit individual taste. Decorate with chopped walnuts or almonds. You can freeze the unbaked dough for 6 weeks, baked cookies for 3 months.

½ cup butter, at room
    temperature
½ cup honey
1 egg, beaten
¾ cup molasses
4 cups whole wheat pastry
    flour
1 teaspoon baking soda
¼ teaspoon ground anise,
    optional
½ teaspoon ground
    cloves

½ teaspoon ground cloves
½ teaspoon ground nutmeg
1 teaspoon ground
    cinnamon
½ teaspoon ground
    cardamom or
    coriander
3 tablespoons sour milk or
    buttermilk
¼ cup chopped nuts
    chopped walnuts or
    almonds for decoration

In a large bowl, cream butter. Add honey and beat until combined. Stir in egg and molasses. In another bowl, combine flour, baking soda, anise, cloves, nutmeg, cinnamon and cardamom. Add alternately with the milk to the molasses mixture. Stir in ¼ cup nuts.

Store, covered, overnight in the refrigerator. Drop dough by teaspoons onto ungreased cookie sheet. Decorate with chopped walnuts. Bake at 375°F for 10 to 15 minutes. The shorter baking time results in a softer cookie.

Frozen dough should thaw in the refrigerator for 6 to 8 hours, then proceed as directed. Frozen baked cookies may be thawed, separated, for 15 minutes at room temperature.

YIELD: 48 to 60 cookies

LEMON BUTTER COOKIES

Dough or baked cookies can be frozen for 3 months.

1 cup butter, at room
  temperature
¾ teaspoon lemon extract
⅓ cup honey
2 egg yolks

grated peel of ½ lemon
1½ cups whole wheat pastry
  flour
36–48 pecan halves

In a large bowl, using an electric mixer, cream butter and lemon extract until fluffy. Gradually add honey, egg yolks and lemon peel and beat well. Add flour to the butter mixture and mix thoroughly.

Cover and refrigerate for 2 hours for easier handling. Form dough into two rolls, 1½ inches in diameter. If freezing for later use, wrap each roll in plastic wrap and plastic freezer bags or wrap in aluminum foil.

With a sharp knife, slice rolls into ¼-inch slices and place on lightly oiled baking sheets, allowing 2 inches between cookies. Press a pecan half into the center of each cookie. Bake in 375°F oven for 7 minutes or until lightly browned.

To bake frozen dough, remove from freezer and let stand at room temperature for 30 minutes. Slice and bake as directed.

Frozen baked cookies should be thawed, unwrapped, at room temperature for 15 minutes.

YIELD: 36 to 48 cookies

VARIATIONS: Substitute vanilla extract for lemon extract and add ½ cup small carob chips or finely chopped nuts.

ALMOND CRESCENTS

Freeze for 4 months.

1 cup butter, at room
  temperature
⅓ cup honey
2 cups whole wheat pastry
  flour

1½ cups ground almonds
1 teaspoon vanilla extract

In a large bowl, cream butter. Gradually add 2 tablespoons of the honey, flour, 1 cup of the almonds and vanilla. Mix thoroughly. Shape dough into a long bar with flat bottom and rounded top and sides, approximately 2 inches across. Freeze until firm, about 30 minutes. Or, at this point, the dough can be frozen up to 6 weeks. To thaw, let stand 1 hour at room temperature.

Cut bar into ½-inch-thick slices and bend in ends slightly to form a crescent.

Bake in 350°F oven on ungreased cookie sheets for 30 to 35 minutes, until lightly browned.

Warm remaining honey and lightly brush honey on each warm cookie and immediately roll in remaining ground almonds. Serve, or wrap and freeze.

Thaw frozen cookies, unwrapped, at room temperature for 15 minutes.

YIELD: 50 cookies

After the dough has thawed, cut ½-inch-thick slices, then gently form the slices into crescents and place on a cookie sheet.

PIZZELLE

These thin Italian cookies can be used to make delicious ice cream sandwiches. If shaped while hot, they can also be made into ice cream cones or custard cups. Pizzelle can be frozen for 4 months.

3 eggs, lightly beaten	1 teaspoon baking powder
6 tablespoons honey	2 teaspoons vanilla extract
¾ cup melted butter	1 teaspoon aniseeds,
2 cups whole wheat pastry flour	optional

Preheat pizzelle grid until hot, about 10 minutes.

In a medium-size bowl, whisk together eggs and honey. Beat in butter. Gently whisk in flour and baking powder. Stir in vanilla and aniseeds.

Drop 1 rounded tablespoon of batter in the center of each grid. Quickly close lid and clip handles together. Cook until steaming stops, about 1 minute.

Remove with fork and immediately shape into cones or tart cups or cool on wire rack.

Use immediately or freeze cooled pizzelle in freezer bags. Thaw before using, at room temperature for about 15 minutes.

YIELD: 20 pizzelle

NOTE: Pizzelle can only be shaped into cones or custard cups *before* freezing.

VARIATION: Omit vanilla and aniseeds and replace with 2 teaspoons lemon juice and 1 teaspoon grated lemon zest.

CHEESECAKE COOKIES

Savor the joy of cheesecake without overindulging. These cookies will freeze for 3 months.

Topping

2 cups blueberries, pitted sour cherries or strawberries, partially thawed if frozen

¼ cup honey
1½ tablespoons Mochiko rice flour

Crust

½ cup wheat germ
½ cup ground nuts

3 tablespoons melted butter

Filling

1 egg
8 ounces cream cheese, at room temperature

¼ cup honey
1 teaspoon vanilla extract

To make topping, combine blueberries, honey and rice flour in a medium-size saucepan. Stir over medium heat. Cook slowly until mixture thickens. Cool and refrigerate until ready to use.

To make crust, mix together wheat germ, nuts and butter to make a crumbly mixture. Line 12 large or 24 small muffin cups with cupcake papers. Press the wheat germ mixture into the bottoms and halfway up the sides of the cups.

To make filling, blend egg, cream cheese, honey and vanilla until smooth. Fill cups halfway with this mixture. Bake at 350°F, 10 minutes for the small cookies, 15 minutes for the large ones. Allow to cool, then spoon some topping over each. The cookies can be frozen at this point, in single layers.

Thaw frozen cookies unwrapped at room temperature for 3 hours or wrapped in the refrigerator overnight.

YIELD: 12 large or 24 small cookies

VARIATION: Instead of fruit, top with chopped nuts before baking.

GINGERBREAD WAFFLES

You can freeze these waffles for 2 months.

¼ cup butter, at room
     temperature
¼ cup honey
½ cup molasses
2 eggs, separated, at room
     temperature
1 cup milk
2 cups sifted whole wheat
     pastry flour

1½ teaspoons baking powder
1 teaspoon ground
     cinnamon
1 teaspoon ground ginger
¼ teaspoon ground
     allspice

In a large bowl, beat butter, honey and molasses together until well blended. Beat in egg yolks and milk.

In a medium-size bowl, stir together flour, baking powder, cinnamon, ginger and allspice.

Stir dry ingredients into wet mixture just enough to moisten all ingredients.

In a small bowl, beat egg whites until stiff. Gently fold whites into batter.

Bake in preheated Belgium or regular waffle iron until lightly browned, about 5 to 7 minutes. Cool and wrap for freezing, or serve with ice cream or whipped cream.

To reheat, toast frozen waffles in a toaster or heat in a 350°F oven until warm, about 15 minutes.

YIELD: 4 Belgium waffles, 6 regular waffles

BELGIUM COCONUT WAFFLES

These delicious waffles can be frozen for 2 months.

1½ cups whole wheat pastry
     flour
2 tablespoons potato
     starch
2 teaspoons baking powder
¾ cup shredded coconut

2 tablespoons maple syrup
2 tablespoons honey
3 eggs, separated, at
     room temperature
4 tablespoons vegetable oil
1½ cups milk

In a large bowl, stir together flour, potato starch, baking powder and coconut.

In a small bowl, whisk together maple syrup, honey and egg yolks. Whisk in the oil and milk.

Make a hole in the center of the dry ingredients. Pour liquid mixture into the hole. Stir just enough to moisten all ingredients.

In a medium-size bowl, beat egg whites until stiff. Gently fold whites into batter.

Bake batter in preheated Belgium or regular waffle iron until golden brown, about 6 to 8 minutes.

Cool and wrap for freezing or serve with ice cream, whipped cream, nuts and/or toasted coconut.

To reheat, toast frozen waffles in a toaster or heat in a 350°F oven until warm, about 15 minutes.

YIELD: 6 Belgium waffles, 8 regular waffles

BASIC WHOLE WHEAT PASTRY CRUST

Can be frozen for 2 months.

6 tablespoons cold butter	5–6 tablespoons yogurt
1½ cups whole wheat pastry flour	

With two knives or a pastry blender, cut butter into flour until mixture resembles coarse crumbs.

Add yogurt, 1 tablespoon at a time, and stir until dough holds together.

Roll dough between two sheets of wax paper until about ⅛ inch thick and 1 inch larger than a 9-inch pie plate. Fit into pie plate and cut off excess dough, leaving a 1-inch overhang. Tuck overhang under dough around rim and crimp or flute with fingers. Crust can be placed in plastic freezer bag and frozen.

YIELD: one 9-inch pie crust, twelve miniature shells

VARIATION: For use in quiches and other savory pies, mix 1 tablespoon Parmesan cheese with flour before combining with the butter.

APPLE STRUDEL

The filling for this strudel can be frozen up to 3 months. Baked strudel may also be frozen for 3 months.

4 tart apples (such as Granny Smith), peeled and thinly sliced, or 3 cups slices, thawed if frozen	½ cup chopped nuts
	¼ cup butter
	½ cup heavy cream
	¼ cup honey
	16 phyllo sheets
grated peel and juice of 1 lemon	¾ cup melted butter, approximately
½ cup raisins	

Place apples in a large frying pan with lemon peel and juice, raisins, nuts and butter. Sauté over medium heat for about 5 minutes, until apples start to soften.

Add cream and honey. Cook, stirring often to prevent sticking, until sauce is very thick and apples are soft but not mushy. Set aside to cool. At this point, the filling may be frozen. Thaw it completely before using it.

Place one phyllo sheet on a damp (not wet) tea towel on the work surface. Using a pastry brush, very carefully and thoroughly coat the sheet with melted butter. Place another sheet of phyllo over the first and repeat the buttering procedure. Repeat until four sheets of dough have been used.

Take one-quarter of the apple filling (about ¾ cup) and place along one short edge of the dough, leaving about an inch border around edges. Using the towel to help, roll dough to enclose filling. Tuck edges under to seal in filling. Carefully transfer to a buttered baking sheet. Brush roll thoroughly with butter.

Repeat entire procedure with filling and phyllo to make three more rolls. Be sure to butter each one after transferring it to the baking sheet so the dough won't dry out.

Bake at 350°F for 25 to 30 minutes, until rolls are golden. Some of the rolls may split as they bake.

Wrap now and freeze or serve while still warm. Cut carefully with a serrated knife. (Cold strudel can also be reheated.)

When ready to serve frozen strudel, reheat by baking unthawed strudel at 400°F for 20 to 25 minutes.

YIELD: 16 to 24 servings, 4 strudels

BASIC TURNOVER DOUGH

This dough can be used to prepare stuffed turnovers. Fillings should be dry and cool when you use them. You will need about 2¼ cups of filling for one recipe of turnovers. Do not freeze dough. Assembled turnovers, however, can be frozen for 6 months.

1-1¼ cups whole wheat flour
  2 tablespoons butter, at
      room temperature
  1 egg

¼ cup warm water
  butter for cooking,
   optional

Place ¾ cup of the flour in a medium-size bowl. With a fork, work butter and egg into the flour. Add water and stir to form a dough.

Turn dough onto a counter with remaining flour and knead for about 5 minutes, until dough is smooth and soft. Shape dough into a ball, place in a bowl, loosely cover with plastic wrap and chill for at least 30 minutes before proceeding. The dough can be kept up to 2 days in the refrigerator.

Divide dough in half. Pat each half into a 5-inch square on a lightly floured counter. Roll dough out to a rectangle 15 inches by 10 inches, flouring as needed.

To make turnovers, cut dough into twelve 5-inch squares. Spread 3 tablespoons of filling over half of each square, diagonally, leaving ½ inch of the margin bare. Moisten margin with water and fold dough over filling. Pat gently toward outside edge to flatten filling and eliminate air bubbles. Crimp edges shut with fingers or a fork. Repeat with remaining dough. The turnovers can be frozen at this point.

Cook in butter over medium-low heat, in a loosely covered skillet, turning to lightly brown both sides. Replenish butter as needed.

Frozen turnovers can be placed directly in the skillet, but cooking time will be a little longer.

YIELD: 4 to 6 servings, 12 turnovers

NOTE: Cabbage Filling (page 360) and Palak Parotha Filling (page 359) are two good choices to fill these turnovers.

CHAPTER 13

FAST FOODS
AND PARTY FOODS

It is one of the great virtues of owning a freezer that it can provide food for you when you least want to provide it for yourself. The foods you prepare when you have the time to cook, or when you're feeling particularly experimental or creative in the kitchen, can wait in the freezer for those times when you have neither the time nor the desire to cook.

The freezer is the repository for fast food in its truest sense. This is fast food that is also good food. It may indeed have been long and rather elaborate in the making, but it will be prompt and trouble-free in the serving. Use your own fast food when you want to stave off the hunger of a small child who can't quite wait until dinner is served. Use it when you don't want to cook dinner after a hard day's work. When you have a dinner party to arrange, prepare the fixings ahead of time, so your preparations don't leave you too exhausted to enjoy the fun. There you'll have it, fast food. But it will be fast food that you know is made from whole, natural foods and top-quality ingredients. Your fast food is good food.

The recipes in this chapter are all fast food of a kind that you can serve with the same confidence with which you would serve food that you had made from scratch on the same day. The pizzas, nachos, enchiladas and spring rolls all satisfy your concern with healthful food at the same time that they satisfy any family's fundamental craving for goodies on demand. There's no better antidote to eating food that is full of empty calories than to have food on hand that is not just more nutritious, but more delicious as well.

While we are redefining fast food as the good food you can serve quickly from the freezer, bear in mind that it can also be very uncomplicated food. For instance, sometimes fast food can start with the imaginative use of half a cup of frozen vegetables. Within minutes, you can steam that quantity of broccoli or spinach, chop it, and toss it with some olive oil heated and accented with garlic. Introduce this quick sauce to some pasta, its natural ally, add some basil or parsley, a couple of grindings of black pepper and a little Parmesan, then serve. That's fast food, too, when you use your best cooking instincts to create quickly something from the stock on hand in your freezer.

Fast Food for One

Freezing makes a lot of sense if you live alone. You probably don't want to spend a long time preparing a single portion of a dish because it seems too much effort for too little result. But try cooking six times the quantity you can consume at one sitting, then divide it into single portions and freeze them.

Buy frozen vegetables in big bags, if you live alone. They'll be more nutritious than fresh vegetables left uneaten in the refrigerator, and more flexible than small boxes of frozen vegetables that you may not be able to finish once you've opened them. You can cook a portion from a large bag of vegetables, and keep the rest, tightly fastened, in the freezer.

If you buy bread, leave out half a loaf to eat fresh and keep the other half in the freezer. Buy large packages of chicken parts, chops and ground meat, repackage the contents in small portions and freeze them. Then, even though you cook for one, you can take advantage of cheaper bulk purchases when food is on sale. Wrap the single portions in foil, freeze, then use the same foil to line the baking or broiling pan. It will save on washing up time later. Fast food, fast cleanup.

Freezing Party Food

Use your freezer to give yourself the gift of time. When you have a party to prepare for, choose a menu of dishes that freeze well, and cook as much as you can ahead of time. The day of your party is the time to prepare the fresh components of the feast, including salads and dressings. But the soup, the bread, the main course, the vegetables and the desserts can all be food from the freezer. This book is full of recipes that can be cooked in advance in multiple quantities, frozen and served for a special occasion.

Plan ahead of time a sequence of removing the food from the freezer, and thawing and reheating it. Organize how far ahead to take out foods that need thawing, how far ahead to begin heating up foods that need to go into the oven. Consider whether more than one food will need time in the oven just before the meal is served. If so, will the different dishes need to warm at different temperatures? Planning is clearly necessary, but it is a lot less taxing to plan thawing times than it is to undertake a great deal of cooking on the same day that you plan to entertain.

When you prepare the foods for the party, keep these points in mind. Freezing and reheating softens many foods. For dishes in which texture is an important concern and overcooking would ruin the quality (such as pasta, for example), it is wise to slightly undercook the food before you freeze it. Chill the cooked food very thoroughly in a pan of ice water. The cold will both stop the cooking and take the entire casserole down to a temperature cold enough to resist the formation of large ice crystals in the freezer. Label the food right on the freezer tape with the baking or reheating time it will need, so you can be sure you'll know it when you need to.

Make mouth-watering canapes with rich pastry and fill them with fish or meat or poultry or cheese. Slide them into a 400°F oven right from the freezer, then serve them piping hot.

There are lots of recipes in this chapter and throughout the book that will keep a festive crowd happy. But here are some additional suggestions to consider in using your freezer as a source of food for parties and special occasions. First, consider making party food from your favorite everyday breads. Bread freezes so well, cuts up into so many shapes and sizes and can enclose so many interesting fillings, that it's a natural to freeze and use for party sandwiches, canapes and hors d'oeuvres.

First chill a whole loaf of bread, or as many loaves as you need, in order to make it easier to cut. If you use day-old bread, chilling won't be necessary. Using a knife that is both sharp and hot, cut thin slices from the loaf. You choose the direction—horizontal, vertical, the long way, the short way. Spread each slice with soft, not melted, butter. It will provide a barrier that prevents the filling from making the bread soggy. Either freeze a congenial spread separately or spread it on the bread before freezing.

Ingredients for spreads that work in the freezer include poultry, fish, ground meat, dried fruits, nuts, olives, peanut butter, cheese, anchovies, avocados and spreads made with egg yolk. Food that won't work for frozen canapes include mayonnaise, because it will separate out; egg whites, because they will get rubbery; and tomatoes, which will get mushy. Crisp salad greens, radishes, cucumbers and celery will all go limp because they are by nature very watery. If you are using a dressing to bind your ingredients together, test it to see if it freezes well first by leaving a small quantity in a freezer container in the freezer overnight.

If you make canapes in small batches, one or two kinds at a time, you can freeze them over the course of several days. The filling should always be at room temperature when you spread it on the bread to avoid having it tear the thin slices.

Either tray-freeze the canapes first, or wrap those with like fillings together and freeze them. Separate the layers of canapes on cookie sheets or cardboard covered in freezer wrap. Stack layers of canapes no more than two deep, and then overwrap them with foil or plastic wrap, excluding as much air as possible. Research has shown that canapes frozen when in contact with the sides of the freezer do not freeze uniformly. When you place freezer packages of canapes in cold storage, keep them away from the interior walls. Otherwise, ice crystals will form at the point of contact between the walls and the bread, making it very soggy.

Keep canapes in the freezer for no longer than two weeks. Thaw them, still wrapped, in the refrigerator for an hour or two. The bread will dry out quickly once thawed and is best kept wrapped until you are ready to serve it. The food will be none the worse for having been in the freezer, and you will be very much the calmer for having it ready when you need it.

Party Ice

In addition to serving as a storehouse for food, your freezer can also be the means of creating decorative effects. Party ice, for one, can float delicately in the punches and fruit juices (some from your own freezer, of course) that you serve guests.

Here are several suggestions for making pretty ice cubes. Place a fresh mint leaf in each section of an ice cube tray. Pour mint tea over the leaves and freeze. Keep it wrapped and ready to add to a green punch. Or, try adding fresh or frozen fruit to an ice cube tray, then pour juice over the fruit, and freeze. If you use strawberries or raspberries as the fruit, use apple, cherry or cranberry juice to pour over them. They'll turn your punch pink.

Arrange edible flowers like violets, nasturtiums and rose petals in a ring mold. Cover them with water or juice and freeze. The effect will be both pretty and surprising when you add it to a punch bowl. The larger the mold you use, the longer it will take to thaw. You might also fill a mold with red, hibiscus-based tea and freeze it, to make a dramatic presentation in the middle of a punch bowl.

PINE NUT AND BASIL
CHEESE BALL

Can be frozen for 3 months.

1 pound longhorn cheese, or any other hard, mild cheese
8 ounces cream cheese, at room temperature
1 tablespoon lemon juice

¼ cup lightly packed fresh basil, approximately
1 clove garlic, crushed
grated nutmeg, to taste
2-3 tablespoons pine nuts

Shred longhorn cheese into a medium-size bowl. Stir in the cream cheese until well combined.

Pour lemon juice into a small bowl. Mince basil and stir into the lemon juice.

Add basil mixture, garlic and nutmeg to the cheese mixture and shape into a ball.

Press pine nuts into the surface, making a spiral pattern.

Chill 1 hour and serve, or wrap in wax paper, place in a freezer bag and freeze.

Thaw the frozen cheese ball, wrapped, overnight in the refrigerator.

YIELD: 1½-pound cheese ball

HERBED CHEESE BALL

Can be frozen for 3 months.

1 pound longhorn cheese
8 ounces cream cheese at room temperature
3 tablespoons chopped fresh dillweed

¼ cup sunflower seeds, toasted (page 54)
¼-½ teaspoon cayenne pepper
2 teaspoons paprika

Shred longhorn cheese into a bowl. Stir in the cream cheese and combine thoroughly. Add dillweed, sunflower seeds and cayenne to the cheeses. Shape mixture into a ball and roll in paprika.

Chill 1 hour and serve, or wrap in wax paper, place in a freezer bag and freeze.

Thaw the frozen cheese ball, wrapped, overnight in the refrigerator.

YIELD: 1½-pound cheese ball

STUFFED MUSHROOMS FLORENTINE

This favorite appetizer can be prepared up to 1 month in advance and frozen to be baked on short notice.

1½ cups chopped lightly
    cooked spinach,
    thawed if frozen,
    well drained
4 ounces cream cheese, at
    room temperature
½ teaspoon fresh thyme
    leaves, or ¼ teaspoon
    dried thyme

12-16 large mushrooms, stems
    removed
1½ tablespoons vegetable
    oil
½ cup bread crumbs

If using freshly cooked spinach, cool slightly. Place spinach on a cutting board and mince with a stainless steel knife. Place spinach in a bowl and stir in the cream cheese and thyme.

Sauté mushrooms in oil over high heat until mushrooms are heated through, but not cooked.

Remove mushrooms from heat (and cool if you plan to freeze the dish). Divide spinach mixture among the caps, piling it high. Roll each mushroom in bread crumbs.

To freeze, place on a baking sheet and freeze. When solid, transfer to freezer bag.

To bake immediately, place mushrooms in casserole and bake at 375°F for about 15 minutes, until they are heated through and lightly browned.

To bake from the freezer, keep covered and add 15 to 20 minutes to baking time. Remove cover for the last 10 minutes so the tops can brown. If tops become too brown, cover again.

YIELD: 4 servings

SPRING ROLLS

Varying the fillings for spring rolls (similar to egg rolls) is easy. The only rule is that the filling must be very dry and cooled to room temperature. Have 3 cups of filling for twelve fat spring rolls. Freeze spring rolls after a brief frying, then reheat and finish cooking in the oven. They can be frozen for 4 months.

3 dried shiitake mushrooms	1 cup shredded carrots
½ cup lukewarm water	½ cup chopped scallions,
1 small head celery	including tender green
cabbage, or 2 cups	tops
chopped assorted	2 cloves garlic, minced
vegetables	1 teaspoon minced fresh
2 tablespoons vegetable	gingerroot
oil plus oil for shallow	2 teaspoons soy sauce
frying	1 tablespoon whole wheat
¼ pound boneless chicken	flour
or pork, minced	2 tablespoons water
8 mushrooms, thinly sliced	12 large egg roll skins

Soak shiitake mushrooms in the lukewarm water for 30 minutes. Remove mushrooms (but save the flavorful water for soup), discard hard stems and cut mushrooms into ½-inch pieces.

Remove thick stalks from the cabbage leaves. Cut stalks, across the fibers at a slight angle, into thin slices. Chop leaves into ½-inch squares (keep separate from stalks).

Place wok over high heat; add 1 tablespoon oil. When hot, stir-fry chicken for about 2 minutes or until opaque. Remove from wok with a slotted spoon and cool slightly.

Add remaining tablespoon of oil to wok. When hot, add cabbage stalks and fresh mushrooms. Stir-fry 1 minute, then, stirring after each addition, add carrots, shiitake mushrooms, scallions, garlic and gingerroot.

Add cabbage leaves; stir until completely wilted. It is important that the filling be dry. If a bit more cooking will evaporate accumulated liquid, cook further. If not, drain everything in a colander to remove excess liquid. Add reserved meat and soy sauce. Cool.

Make a paste to seal the loose ends of the egg roll skins by mixing flour with water.

Spread six egg roll skins on a counter. Fill according to diagram, using ¼ cup filling for each. Seal last edge with flour paste. Repeat with remaining skins.

Put enough oil in a clean wok so that two spring rolls can be shallow-fried at one time. Heat oil until a scallion ring dropped in it dances in circles.

Fry spring rolls two at a time, turning with tongs to brown both sides. Drain on paper towels. Freeze at this point, or transfer to a dish and bake, uncovered, in a 350°F oven for at least 10 minutes before serving. Serve hot with Chinese hot mustard.

To prepare frozen spring rolls, bake directly from freezer, uncovered, in 350°F oven until sizzling hot, about 35 minutes.

YIELD: 12 spring rolls

VARIATIONS: Tofu or shrimp can be substituted for the chicken or pork.

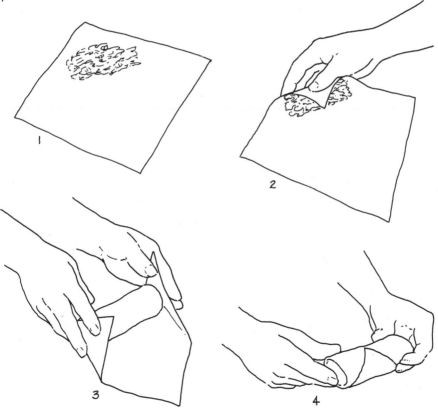

To make the spring roll, hold the square of dough with one corner facing you. 1. Place the filling toward the center, just above the corner. 2. Fold the corner facing you up and over the filling and roll slightly forward. 3. and 4. Fold the side flaps in toward the middle and continue rolling, sealing the last flap with flour paste.

CLAMS CASINO

Can be frozen for 4 months.

36 littleneck or cherrystone clams
3 shallots
½ cup chopped fresh basil
1 teaspoon fresh oregano or ½ teaspoon dried oregano
3 cloves garlic
1 small tomato, cored and halved
4 small mushrooms
2 tablespoons grated Parmesan cheese
½ cup bread crumbs freshly ground black pepper, to taste
Tabasco sauce, to taste, optional
¼ cup grated mozzarella cheese
¼ cup olive oil
2 tablespoons white wine vinegar
¼ cup apple juice

Open clams and rinse thoroughly under cold water. Discard the top shell and loosen the clam on the bottom half.

In a medium-size bowl, combine shallots, basil, oregano, garlic, tomatoes, mushrooms, Parmesan cheese, bread crumbs, pepper and Tabasco sauce.

Transfer this mixture to a food processor or blender. Process to a medium puree, scraping the sides of the container when necessary.

Divide this mixture among all the clams. Top each clam with mozzarella cheese. Place shells in a large baking dish or rimmed baking sheet.

At this point, the clams may be frozen for later use by covering with wax paper and foil.

When ready to bake, mix together the oil, vinegar and apple juice. Gently pour over the clams (or brush over the clams if frozen).

Bake at 400°F until golden brown, 10 to 20 minutes for fresh clams, 15 to 25 minutes for frozen.

YIELD: 36 clams

ITALIAN MEATBALLS

You can freeze these meatballs for 6 months.

4 slices whole wheat bread
1 cup water
1 pound lean ground beef,
  thawed if frozen
2 eggs
½ cup grated Parmesan or
  Romano cheese
2 tablespoons chopped
  fresh parsley
2 cloves garlic, minced
2 teaspoons fresh oregano
  or 1 teaspoon dried
  oregano

1 teaspoon fresh thyme
  leaves, or ½ teaspoon
  dried thyme
½ teaspoon fresh rosemary,
  or ¼ teaspoon crushed
  dried rosemary
  ground black pepper to
  taste
2 tablespoons olive oil
1 recipe Italian Tomato
  Sauce (page 56),
  thawed if frozen

Soak bread in water for 1 minute. Squeeze out moisture.

In a large bowl, combine bread, beef, eggs, cheese, parsley, garlic, oregano, thyme, rosemary and pepper. Mix well. Shape into balls 1 to 1½ inches in diameter.

Heat oil in a large skillet. Over medium heat, slowly brown the meatballs. Add tomato sauce and simmer, covered, for 30 minutes.

To freeze, place cooled meatballs and sauce into freezer bags or containers.

To reheat frozen meatballs, place while frozen in a baking pan, cover and bake at 350°F until heated through, about 35 to 40 minutes. Check periodically to make sure the meatballs are not drying out. If they appear dry, add a little water to the pan.

YIELD: 40 meatballs

SWEET-SOUR COCKTAIL MEATBALLS

Can be frozen for 6 months.

1¼ pounds lean ground beef,
     thawed if frozen
1   cup cooked brown rice
¼   cup finely chopped
     onions
¼   cup whole wheat bread
     crumbs
2   teaspoons soy sauce
⅛   teaspoon ground black
     pepper
1   tablespoon vegetable oil
1½ cups Beef Stock
     (page 71), thawed if
     frozen

2   tablespoons vinegar
2   tablespoons honey
1   teaspoon finely chopped
     fresh basil, or
     ½ teaspoon dried basil
¾   teaspoon ground ginger
3   tablespoons catsup
2   teaspoons cornstarch
2   tablespoons water

In a large bowl, mix together beef, rice, onions, bread crumbs, 1 teaspoon of the soy sauce and pepper. Form mixture into about 30 walnut-size balls.

Heat oil in a large skillet over medium heat and lightly brown meatballs on all sides. Turn them carefully, with a metal spatula, to avoid breaking. Drain fat from pan.

Combine beef stock, vinegar, honey, basil, ginger, remaining soy sauce and catsup. Add to meatballs in pan. Cover and simmer for 20 minutes, or until meatballs are cooked through.

Carefully remove meatballs from pan with a slotted spoon and set aside. Mix cornstarch and water. Add to sauce in pan and stir constantly until thickened and bubbly. Return meatballs to pan and stir gently to cover all the meatballs with sauce. The meatballs can be frozen at this point, or heated through before serving.

To make frozen meatballs, transfer to an ovenproof casserole and bake, covered, in a 350°F oven until hot and bubbly, 35 to 40 minutes.

YIELD: 40 meatballs

CHEESY SAUSAGE BALLS

An easy hot and spicy appetizer. Make these ahead of time for your next party. These are not low in calories, but they can be made so by using a low-fat cheese and a lean sausage. Can be frozen for 3 months.

1¼ cups whole wheat flour
1 pound bulk hot sausage,
    thawed if frozen
1 pound cheddar cheese,
    shredded

¼ cup buttermilk, thawed
    if frozen
2 eggs, beaten

In a large bowl, combine flour, sausage and cheese. Stir in the buttermilk and eggs and mix thoroughly. Form into balls ¾ inch in diameter.

To freeze for later use, place meatballs on a rimmed baking sheet. Place sheet in freezer until balls are firm, then transfer them to freezer bags until needed.

To bake, place balls on an oiled broiler pan. Broil fresh balls for 12 to 15 minutes, frozen balls for 15 to 20 minutes, until lightly browned. Drain on paper towels or brown paper bags. Serve hot.

YIELD: about 60 balls

FRAGRANT ARTICHOKE HEARTS

This delightful appetizer can be frozen for 3 months.

1½ cups artichoke hearts,
  thawed if frozen
1 cup chopped tomatoes,
  thawed if frozen
1 teaspoon minced fresh
  oregano, or ½ teaspoon
  dried oregano

1-2 cloves garlic, minced
2 tablespoons lemon juice
  freshly ground black
  pepper, to taste
1 tablespoon grated
  Parmesan cheese
¼ cup olive oil

Place artichokes in a lightly greased 1-quart ovenproof casserole dish.

In a small bowl, stir together the tomatoes, oregano, garlic, lemon juice and pepper. Spread mixture over artichokes.

Sprinkle with cheese and pour oil over top.

Cover and freeze, or bake at 350°F until bubbling, about 25 to 30 minutes.

To bake frozen casserole, keep covered and increase baking time to 30 to 40 minutes.

YIELD: 4 servings

BROCCOLI TIMBALES

An excellent way to use frozen broccoli, this dish is especially good served with a cheese sauce or a fresh tomato sauce.

Underbaked timbales can be frozen up to 4 months.

2 cups cooked frozen
  broccoli, thawed
4 eggs
1 cup milk

2 tablespoons whole wheat
  flour
½ teaspoon dried oregano
⅛ teaspoon cayenne pepper

Puree broccoli in blender or food processor. Add eggs, milk, flour, oregano and cayenne. Process until thoroughly mixed.

Butter four single-serving ramekins or baking dishes. Divide broccoli mixture among them and bake at 350°F for 25 to 35 minutes.

Timbales are done as soon as they appear fairly firm when lightly shaken. If you want to freeze the timbales, remove them from the oven a few minutes earlier—when outside edge begins to set. Cool and freeze.

To serve, run a knife around inside of cups and invert onto serving plates or serve in baking dish.

To reheat frozen timbales, bake, covered, at 350°F until hot, about 35 to 40 minutes.

YIELD: 4 servings

VARIATION: Substitute asparagus for the broccoli.

SPINACH AND FETA PIES

These pies are excellent with frozen spinach or broccoli. They freeze well and are handy in individual-size portions. The pies are extra fast to make because the pie shell is simply walnuts and wheat germ. You can freeze the pies for 6 months.

3 cups chopped cooked spinach, thawed if frozen, squeezed dry	4 cloves garlic, minced
	½ teaspoon pepper
	2 teaspoons dried oregano
2 cups crumbled feta cheese	¾ teaspoon ground cinnamon
8 eggs, beaten	⅓ cup ground walnuts
1 cup yogurt	⅓ cup toasted wheat germ

In a large bowl, combine spinach, cheese, eggs, yogurt, garlic, pepper, oregano and cinnamon, using a fork.

In a separate bowl, combine walnuts and wheat germ.

Liberally butter two 9-inch pie pans or eight individual 4½-inch pans, then coat evenly with walnut mixture.

Pour spinach-feta mixture into prepared pans.

Bake at 400°F until barely set, about 30 to 35 minutes for the 9-inch pies, 25 to 30 minutes for the smaller pies. The pies may be frozen at this point for reheating later. Or, continue baking an additional 5 to 10 minutes for serving immediately.

To reheat frozen pies, cover loosely with foil and bake at 350°F for 30 to 45 minutes for a small pie, 40 to 60 minutes for a large pie.

YIELD: two 9-inch pies or eight smaller pies

ARABIC SPINACH PIES

This recipe is a hybrid of Greek and Syrian spinach pies, really more of a turnover. The obvious freezer approach here is to freeze the assembled and baked pies for later reheating. But for the freshest flavor, you may choose to prepare the filling from frozen ingredients and encase it in fresh dough, perhaps a favorite freezer dough. Baked pies can be frozen for up to 3 months, unbaked pies for 2 months.

Filling

½	cup diced onions	1	cup crumbled feta cheese
2	cloves garlic, minced	1	egg
1	tablespoon olive oil		dash of cayenne pepper
15	fresh mint leaves, finely chopped		
½	teaspoon dried oregano		
¾	pound coarsely chopped fresh spinach or 2 cups chopped frozen spinach, thawed and drained		

Dough

1	tablespoon active dry yeast	2	tablespoons vegetable oil
¼	cup warm water	1	tablespoon honey
¾	cup milk	1	egg
		2½–3	cups whole wheat flour
3	tablespoons melted butter		

The filling may be prepared in advance, and should be at room temperature when the dough is stuffed. (The dough will require 2 to 3 hours from beginning to end.)

To prepare the filling, sauté onions and garlic in oil in a 3-quart covered pot over medium heat for 5 minutes or until onions start to soften. Stir in mint and oregano, and cook 1 minute more.

Drain spinach well, then add to the pot. Cover and cook about 5 minutes for fresh spinach, 3 to 5 minutes for frozen. Uncover pot and cook until excess liquid is gone.

Cool to room temperature and mix in cheese, egg and cayenne.

To make the dough, dissolve yeast in the warm water.

Warm milk to 110°F, and mix in oil, honey and egg. Put 2½ cups of the flour into a bowl. Make a well in the flour and pour in the milk and yeast mixtures. Stir well to form a soft dough. If the dough is too sticky, gradually add the remaining ½ cup flour, mixing thoroughly after each addition.

Turn the dough onto a lightly floured surface and knead for about 10 minutes. Try to incorporate as little additional flour as possible. When the dough is smooth and feels lively or elastic, transfer to an oiled bowl. Flip the oiled side of the dough up and cover the bowl with a damp cloth. Set to rise in a warm spot until doubled in size, about 1 hour.

Punch down the dough, knead briefly and divide into ten equal pieces. With oiled hands, shape dough into balls. Place them on lightly buttered baking sheets. If dough is dry, cover loosely with plastic wrap. Let rise again until doubled in size, 20 to 30 minutes. Roll out each ball of dough to a 5-inch circle, flouring lightly if necessary. Return the circles to the baking sheets.

To assemble the pies, lightly spread ¼ cup of the spinach filling over half of each circle, avoiding the outside ½ inch. Moisten edge with water and fold circle to form a half moon. Gently crimp edge with a fork to seal it. Brush each pie with melted butter. The pies may be frozen, unbaked, at this point.

To prepare immediately, bake in a 375°F oven for 20 to 25 minutes until lightly browned. Remove from the oven and brush again with butter. Cool slightly before serving or before wrapping to freeze.

If frozen unbaked, bake as directed above, adding about 25 minutes to baking time. To reheat baked pies, wrap in foil and bake at 325°F until hot, about 35 minutes.

YIELD: 10 small pies

VARIATIONS: Almost anything is possible. You will need 2½ cups of filling. You can replace spinach with chopped cooked broccoli or chard. Or instead of spinach and cheese, use ground lamb or beef sautéed with lots of onions and garlic. For a French flavor, substitute a milder cheese for the feta, such as Swiss or muenster, and use thyme and nutmeg in place of the mint leaves. Or use a rich pie pastry instead of a yeast dough.

SPINACH BALLS

These snacks are much tastier than the popular convenience food recipe they're based on. You can freeze them for 4 months.

2 cups chopped cooked spinach, thawed if frozen, well drained
3 cups bread crumbs
½ cup finely chopped onions
½ cup chopped celery
6 eggs, beaten
¾ cup melted butter
½ cup grated Parmesan cheese
1 or 2 cloves garlic, minced
¼ teaspoon dried thyme
¼ teaspoon ground black pepper
2 drops hot pepper sauce
1 teaspoon dried sage
¼ teaspoon crushed rosemary
¼ teaspoon ground nutmeg
2 teaspoons soy sauce

In a large bowl, mix together all ingredients.

Cover and place in refrigerator at least 1 hour or overnight to firm the mixture for easier rolling. Shape mixture into 1-inch balls. Balls can be frozen at this point for later use.

Bake on buttered baking sheets in a 350°F oven for 20 minutes for fresh balls, 25 minutes for frozen balls.

YIELD: about 70 balls

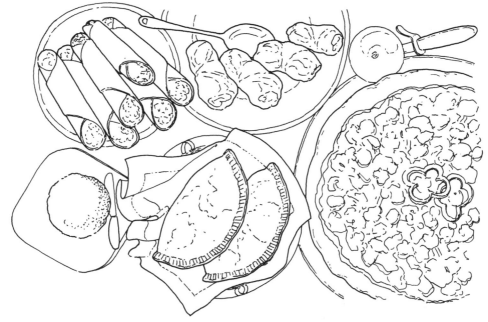

RICE BALLS

Can be frozen for 4 months.

1 cup short-grain brown rice	3 tablespoons grated Parmesan cheese
1¾ cups water	½ teaspoon dried thyme
2 tablespoons butter	1 egg
1 cup minced onions	1 cup bread crumbs
2 cloves garlic, minced	vegetable oil for frying
¼ pound cheddar cheese, shredded	

Place rice and water in a pot and bring to a boil. Cover and cook over medium-low heat until water is absorbed and rice is soft, about 30 minutes, stirring occasionally while the rice is cooking. (If necessary, add more water to keep the rice from sticking to the bottom of the pot.) Stirring makes the rice sticky, which is important to this recipe.

Melt butter in a small skillet. Add onions and garlic and sauté gently until the onions are soft. Do not allow the mixture to brown.

In a medium-size bowl, mix rice, onions and garlic, cheeses, thyme and egg. Roll the mixture 1 tablespoon at a time in the bread crumbs and set aside. The rice balls can be frozen, uncooked, at this point.

Heat ⅛ inch of oil in a skillet over medium heat. Fry balls until they are brown, being careful not to overcrowd the pan, and turning the balls so that all sides are done. If planning to freeze the cooked balls, remove from pan when slightly undercooked. Drain on paper towels. Serve at once or keep in a 300°F oven for up to 15 minutes until serving.

Freeze uncooked or slightly undercooked rice balls on a baking sheet. When solid, transfer to a freezer bag. To reheat undercooked balls, thaw and warm in a 300°F oven for 15 minutes before serving. Raw rice balls can be thawed and fried as above.

YIELD: 40 to 45 balls

ENCHILADAS

Traditionally, enchiladas are made with tortillas that have been quickly fried in hot oil before rolling. To avoid the frying process, soften tortillas in a warm oven for a few minutes. Cover with a damp towel while preparing filling.

The creative cook can devise many variations on this filling. Garnish the enchiladas with sour cream, guacamole and chopped lettuce.

Enchiladas can be frozen (without garnishes) for 3 months.

12 5- to 6-inch tortillas, thawed if frozen	4 cups shredded Monterey Jack or longhorn cheese
½ cup finely chopped onions	4 cups Mexican Coriander Sauce (page 46), thawed if frozen and warmed
2 tablespoons finely chopped sweet green peppers	
1 tablespoon olive oil	

Wrap tortillas in aluminum foil and warm in 350°F oven for several minutes.

In a skillet, sauté chopped onions and peppers in oil until tender, about 3 minutes. Stir onion mixture into 3 cups of the cheese.

Dip tortillas in warm sauce. Place about ¼ cup cheese filling at one end of each tortilla and roll loosely, crepe fashion. Place enchiladas seam-side down in a 9 × 13-inch baking dish. Cover enchiladas with 2 cups of the sauce and top with remaining cheese. Extra sauce can be served on the side.

Bake, uncovered, at 350°F for 20 minutes.

Enchiladas may be frozen, unbaked, covered with sauce and cheese in a covered baking dish. Bake frozen enchiladas, covered, at 350°F for 25 minutes. Uncover and continue baking until hot and bubbly, about 15 minutes more.

YIELD: 6 to 8 servings

VARIATIONS: Substitute 1 pound cooked loose pork sausage for 3 cups of the cheese. Fill enchiladas with sautéed onions, green peppers, sausage and a little sauce. Top with remaining sauce and 1 to 1½ cups shredded cheese.

Or, mix ½ cup sour cream with 2 to 3 cups shredded cooked chicken. Add sautéed onions and green peppers. Top filled enchiladas with sauce and 1 to 1½ cups shredded cheese. Serve with additional sour cream or guacamole.

Or, add 1 cup cooked kidney beans and ½ pound cooked ground beef to 1 cup Mexican Coriander Sauce. Decrease shredded cheese to 2 cups. Fill each enchilada with a little of the sauce-bean mixture, 1 tablespoon shredded cheese and 1 tablespoon chopped onions. Top filled enchiladas with 2 to 3 cups sauce and remaining cheese. Break 6 eggs on top of enchiladas, if desired, and bake until eggs are set.

NACHOS

Do not freeze.

12 frozen 5- to 6-inch corn tortillas, thawed	½ pound Monterey Jack cheese, shredded
½ cup vegetable oil	2 tablespoons chopped jalapeño peppers
8 ounces frozen refried beans, thawed and warmed	

Quarter tortillas. Heat oil in a large skillet until a drop of water bubbles on the surface. Place cut tortillas, a few pieces at a time, in oil and fry until just crisp. Drain.

Place fried tortillas (*tostaditas*) on a large baking sheet (a pizza pan works well). Top with refried beans and sprinkle shredded cheese over the beans.

Broil until cheese bubbles, 3 to 5 minutes. Top with peppers.

Serve with chopped tomatoes, shredded lettuce, sour cream and guacamole.

YIELD: 4 to 6 servings, 48 nachos

CHICKEN-VEGETABLE TAMALES

Here's a great way to use green tomatoes. Typically this authentic Mexican specialty is usually reserved for holiday meals and special occasions because it takes a great deal of time to prepare. However, it can easily be prepared ahead of time, cooked and then frozen. When that special occasion arises, steam the number of tamales you need for your meal. When they are heated through, serve with Salsa Cruda and sour cream. Accompany this dish with refried beans or brown rice and you'll have a complete Mexican meal.

The tamale consists of a corn flour dough which is spread on a wrapper made of corn husks. A chicken-vegetable filling is spread on top of the dough and the corn husk is then rolled up and tied to keep the tamale together. Tamales can be frozen for 6 months.

1 package corn husks
(10- or 12-ounce
package)

Corn Masa Dough

1½ cups butter, at room
temperature
4 cups creamed corn,
thawed if frozen
¼ cup honey

4 cups masa harina
2 tablespoons baking
powder
1½ cups milk

Chicken-Vegetable Filling

¾ cup finely chopped
onions
2 cups finely chopped
celery
⅓ cup finely chopped
(seeded) green chili
peppers
½ cup diced green bell
peppers

3 cloves garlic, minced
¼ cup vegetable oil
4 cups shredded cooked
chicken, thawed if
frozen
5 large green tomatoes,
peeled and chopped
(drained if canned)
cayenne pepper, to taste

Salsa Cruda (page 48),
thawed if frozen

sour cream

You may use your own dried garden corn husks, or purchase packaged corn husks from a grocery store carrying Mexican products.

Wash husks to remove any remaining corn silk. Soak husks in water a minimum of 2 to 3 hours to soften. If you are short of time, you may soak them in boiling water for 30 minutes to speed up the process.

When you begin the process of assembling your tamales, take a few husks out of the water at a time and gently squeeze out the water. Next, pat husks with a clean dish towel to remove any excess moisture. They will still feel damp. If they are not soaked long enough, they will be stiff and tear easily, making them difficult to use.

To make the corn masa dough, in a large bowl, beat butter with an electric mixer till creamy. Add the creamed corn and honey and continue beating until mixed well.

Mix the masa harina and baking powder together. Add this to the corn and butter mixture, alternately with the milk. The result will be a light and fluffy masa.

To make the filling, sauté onions, celery, chili peppers, green peppers and garlic in oil for 5 minutes. Add shredded chicken and green tomatoes. Season with cayenne and simmer, covered, for 30 minutes. The filling should be moist but not soupy. Remove cover while simmering, if necessary, to eliminate excess moisture. Remove pan from heat.

To assemble the tamales, use one large or two small corn husks stacked together. Spread 4 to 5 tablespoons of masa dough across the wrapper, starting at one edge and making a rectangle approximately 4 × 5 inches. Leave the other three edges bare for wrapping and tying the ends.

Spread 2 to 3 tablespoons of the chicken-vegetable filling on the masa. The filling can be used hot or cold. Roll up the tamale starting at the filled edge and tie the ends with pieces of corn husks.

Steam tamales on a rack over boiling water for 1¼ to 1½ hours. A canning kettle with a colander can be used. To test, remove one of the tamales, allow to cool about 10 minutes and unwrap. They are done if the dough separates from the husk.

Remove from steamer and allow to cool slightly before serving, or cool completely before wrapping and freezing. To reheat after freezing, thaw 2 hours, then steam again until hot, about 30 minutes.

Serve the tamales in the husk, as part of the fun is unwrapping them. To eat, unwrap the corn husk and discard it. Top with Salsa Cruda and sour cream.

YIELD: about 34 tamales

PIZZA CRUST

This crust is not too thick or too thin, but just right! The dough can be frozen for 3 months.

1 package active dry yeast	1¼ cups lukewarm water
¼ cup warm water	¼ cup chopped onions,
3 cups whole wheat flour	sautéed and cooled
2 tablespoons vegetable oil	slightly, optional

Dissolve yeast in warm water. Mix flour, oil, yeast mixture, lukewarm water and sautéed onions until thoroughly blended. Knead for about 5 minutes until smooth and elastic.

Rub a few drops of oil around the bottom of a large bowl. Place dough in bowl and turn, placing oiled side up. Cover with plastic wrap and let rise in a warm place about 1¼ to 1½ hours, until almost doubled in bulk.

Punch down dough and divide in half. Let rise for 15 minutes. Pizza dough may be wrapped in a freezer bag and frozen. To defrost dough, remove from freezer wrap and let rest until near room temperature. Oil two 12-inch pizza pans. Roll out dough to size of pizza pan. Place crust in pan. Fill with your favorite topping or try one of the following recipes.

YIELD: two 12-inch crusts

NOTE: If planning to use this crust for sweet fillings, omit the onions.

GREEK PIZZA

This pizza can be frozen for 1 month.

2½–3 cups chopped cooked spinach, thawed if frozen, squeezed dry	1 large ripe tomato, chopped
1 unbaked 12-inch Pizza Crust (above)	2 scallions, including green tops, chopped
1½ cups crumbled feta cheese	1 tablespoon olive oil
	¼ teaspoon dried basil
	¼ teaspoon dried oregano

Spread spinach over pizza crust. Sprinkle on cheese. The pizza can be frozen at this point in a plastic freezer bag, or bake at 400°F for 10 minutes, until cheese is slightly melted and crust is lightly browned.

Remove from oven. Sprinkle on tomatoes, scallions, oil, basil and oregano. Return to oven for 5 to 10 minutes.

To bake frozen pizza, put in 400°F oven for 20 minutes, then sprinkle with tomatoes, scallions, oil, basil and oregano and return to oven for another 5 to 10 minutes.

YIELD: one 12-inch pizza

TRADITIONAL ITALIAN
PIZZA TOPPING

Pizza can be assembled with this topping and frozen for 3 months.

1 unbaked 12-inch Pizza
    Crust (page 354)
½ cup Basic Tomato Sauce
    (page 56), thawed if
    frozen
⅓ cup chopped onions
½ cup chopped green
    peppers

1 cup shredded mozzarella
    cheese
½ teaspoon dried oregano
½ teaspoon dried basil
⅛ teaspoon garlic powder
1 cup sliced mushrooms

Over pizza crust, spread tomato sauce, onions, peppers, cheese, oregano, basil and garlic powder. At this point, pizza may be wrapped in a freezer bag and frozen.

To make the pizza immediately, bake in 400°F oven until edges of pizza and crust begin to brown lightly, about 10 minutes. Spread mushrooms over pizza. Turn oven down to 350°F and bake until cheese is browned and bubbly, about 10 minutes.

Bake frozen pizza in a 400°F oven for 30 minutes, then turn down oven to 350°F, add the mushrooms and bake 10 minutes more.

YIELD: one 12-inch pizza

MINIATURE FRENCH PIZZAS

Tiny pizzas, made with pie pastry instead of a yeast dough, are prepared in bakeries and other food stores throughout France. It's both fun and simple to make a variety of toppings by using leftovers from the refrigerator and freezer. The pies can be frozen, unbaked, with their toppings, or can be assembled in the frozen shell right before baking. You can freeze them for 3 months.

Three-inch shallow tart shell tins are ideal for forming the pastry but are not necessary. We've divided this recipe into Bases, which are spread over the pastry, and Toppings, which sit on the Base.

2 recipes whole wheat
    pastry (page 329)
4 cups of chosen Base(s),
    listed below

4 recipes of Topping(s),
    listed below

To shape the pastry, divide dough in half. Roll out to 3/16 inch thick. Cut into circles just larger than the 3-inch tart tins. Press into tins. If not using special tins, cut dough into 4-inch circles and place on baking sheets. Carefully crimp the edges inward to form a shallow lip. Keep pastries cool or freeze until ready to fill.

YIELD: 24 tart shells

Base #1

¾ cup ricotta cheese
¼ cup grated Parmesan
    cheese

1 egg
ground black pepper,
    to taste

Mix together until smooth.

YIELD: 1 cup, enough for 6 tarts

Base #2

1 cup sour cream
2 eggs

¼ teaspoon ground nutmeg

Mix together until smooth.

YIELD: 1 cup, enough for 6 tarts

Base #3

1 cup Basic Tomato Sauce (page 56)	seasonings, to taste
tomato paste, optional	

If sauce is thin, add tomato paste. Sauce should be thick. Season if needed.

YIELD: 1 cup, enough for 6 tarts

Topping #1

6 homemade marinated artichoke hearts, quartered	¼ cup grated Parmesan cheese
	oregano, to taste

Top any of the bases with artichokes. Sprinkle with cheese and oregano.

YIELD: enough for 6 tarts

Topping #2

1 teaspoon caraway seeds	paprika
¾ cup shredded Swiss cheese	

Top ricotta base (Base #1) with caraway seeds, then Swiss cheese. Garnish with paprika.

YIELD: enough for 6 tarts

Topping #3

2 scallions, white part only, sliced	2 tablespoons grated Parmesan cheese
6 mushrooms, sliced	6 cherry tomatoes, halved, optional
1 tablespoon butter	

Sauté scallions and mushrooms in butter until tender, about 5 minutes. Use this to top any of the bases. Sprinkle with cheese. If desired, top with tomatoes, cut side up, before baking.

YIELD: enough for 6 tarts

Topping #4

    ½ teaspoon dry mustard   ½ cup shredded cheddar
    2 teaspoons chopped fresh   cheese
      dillweed
    ½ cup tuna, separated
      into chunks

Stir mustard and dillweed into one recipe of the sour cream base (Base #2). Spread on pastry, top with tuna, then cheese. Can be frozen, but is somewhat better fresh.

YIELD: enough for 6 tarts

Topping #5

    ½ cup thinly sliced cooked   ¼ cup shredded Monterey
      spicy sausage   Jack cheese
    ¼ cup finely chopped green
      peppers

Top tomato sauce base (Base #3) with sausage, peppers then cheese.

YIELD: enough for 6 tarts

Topping #6

    3 canned sardines, drained   ¼ cup buttered bread
      and halved   crumbs
    ½ teaspoon dried thyme

Top tomato sauce base (Base #3) with sardines, then thyme. Sprinkle with bread crumbs.

YIELD: enough for 6 tarts

Prepared pizzas can be frozen on baking sheets. When solid, transfer to rigid freezer containers.

    Bake pizzas in a 375°F oven 15 to 20 minutes for fresh pizzas, 30 to 40 minutes if frozen, until cheese melts and/or filling is hot.

PALAK PAROTHA FILLING FOR TURNOVERS

This filling is made with East Indian seasonings. It can be frozen for 4 months.

2 cups lightly cooked, chopped spinach, thawed if frozen, well drained
1 teaspoon ground coriander
½ teaspoon ground cumin
1 teaspoon curry powder
3 cloves garlic, minced
2 tablespoons butter

⅓ cup thinly sliced scallions, including green tops
2 teaspoons finely slivered fresh gingerroot
1 teaspoon lemon juice
1 teaspoon soy sauce
1 cup yogurt

In a large skillet, sauté spinach, coriander, cumin, curry powder and garlic in butter over medium-high heat. Cook until very dry, stirring frequently. Stir in scallions, gingerroot, lemon juice and soy sauce. Cool the mixture quickly by placing it in a bowl set in cold water. If desired, freeze the filling at this point.

Use as described in the recipe for Basic Turnover Dough (page 331). Serve hot with yogurt on the side.

Frozen filling should be thawed in the refrigerator overnight before using.

YIELD: 2¼ cups filling

CABBAGE FILLING FOR TURNOVERS

This recipe makes enough filling for one recipe of Basic Turnover Dough. The filling can be frozen for 9 months.

½ cup diced onions	2 teaspoons caraway seeds
3 cups thinly sliced cabbage, cut into 1-inch lengths, partially thawed if frozen	3 tablespoons butter
	½ cup shredded Jarlsberg cheese
	½ cup sour cream, optional

Sauté onions, cabbage and caraway seeds in butter until all moisture evaporates. Cool and stir in cheese. The filling can be frozen at this point.

Use as directed in the recipe for Basic Turnover Dough (page 331). Serve with sour cream.

To use frozen filling, thaw in refrigerator overnight.

YIELD: about 2½ cups filling

CALZONES

Can be frozen for 4 months.

1 tablespoon active dry yeast	2 cups ricotta cheese
1⅓ cups warm water	4 cloves garlic, minced
3½–4 cups whole wheat flour	1 cup grated Parmesan cheese
4 tablespoons olive oil, approximately	½ cup chopped Italian or curly parsley
1 tablespoon lemon juice	1 teaspoon fresh thyme leaves, or ½ teaspoon dried thyme
1 pound bulk Italian sausage	2 eggs, beaten
¾ cup chopped green peppers	1 pound mozzarella cheese, shredded
1 cup sliced mushrooms	
8 ripe tomatoes, or 1½ cups tomato sauce (page 56), thawed if frozen	

Dissolve yeast in water in a large bowl. Add 3½ cups of the flour, 1 tablespoon of the oil and lemon juice and mix. Knead for 10 minutes or until the dough is shiny and elastic, adding more flour if necessary to make a medium-stiff dough. Oil a large bowl, place ball of dough in the bowl and turn the ball so that all sides are lightly coated with oil. Cover with a damp towel and allow to rise in a warm, draft-free place for about 1½ hours, or until doubled in bulk.

After dough has risen, punch down, turn out onto a lightly floured board and divide into eight equal pieces. Roll each piece out into a circle, about 6 to 7 inches in diameter. Brush each top lightly with olive oil. Set dough circles aside to rise again for about 30 minutes.

Meanwhile, as dough is rising, heat 1 teaspoon of oil in a skillet. Crumble and brown the sausage. Drain and set aside.

Sauté peppers and mushrooms in 1 tablespoon oil until tender. Set aside. If you are using fresh tomatoes, peel, seed and cut them into ¼-inch slices.

Mix ricotta cheese, garlic, Parmesan cheese, parsley, thyme and eggs in a small bowl.

If using fresh tomatoes, arrange slices on half of each dough circle, leaving a ½-inch border on the edges. If using tomato sauce, spread 3 tablespoons on half of each dough circle. Divide sausage, peppers, mushrooms and cheese mixture among the eight circles. Sprinkle each with mozzarella cheese. Brush the uncovered border lightly with water and fold over the other half of the dough circle, pressing edges to seal them. Lightly brush the top of the calzone with olive oil; cut two small slashes to allow steam to escape. Sprinkle baking sheets lightly with cornmeal and place calzones on them.

Bake calzones at 450°F for 15 to 20 minutes, or until the crust is brown and puffed.

If freezing the calzones, bake only 10 to 15 minutes, then cool, wrap in aluminum foil and freeze. To bake, keep wrapped, place in a 350°F oven and bake for 30 to 40 minutes.

YIELD: 8 calzones

INDEX

Page numbers in **boldface** indicate charts. Page numbers in *italics* indicate illustrations.

Rodale Press, Inc., publishes PREVENTION®, the better health magazine.
For information on how to order your subscription,
write to PREVENTION®, Emmaus, PA 18049.